VOLCANOES OF THE SOUTH WIND

A field guide to the Volcanoes and Landscape of Tongariro National Park

Karen Williams

TONGARIRO
NATURAL HISTORY SOCIETY (Inc.)
Friends of Tongariro National Park
World Heritage Site

The Tongariro Natural History Society (Inc.) is a non-profit making organisation co-operating with the Department of Conservation in the interpretive and related visitor-services activities of Tongariro National Park.

TONGARIRO NATURAL HISTORY SOCIETY

The Tongariro Natural History Society was established to promote a wider understanding of the natural processes and human history of Tongariro National Park. At its inaugural meeting in September 1984 the Society was endowed with a substantial Memorial Fund, in memory of:

Keith Maurice Blumhardt
William Edward Cooper
Douglas Neal McKenzie
Derek Ian White
Marie Pauline Williams

who died on Mount Ruapehu in a helicopter accident on 9 December 1982.

The Tongariro Natural History Society funds publications and numerous other worthwhile activities at Tongariro and its work over the past two decades stands as a unique and ongoing memorial to these five people who loved the Park.

The first edition (1985) of 'Volcanoes of the South Wind' symbolises the start of the Tongariro Natural History Society's endeavours. The 2001 edition is a major update and includes a new 16-page section on the 1995-96 Ruapehu eruptions. Although this third edition carries the original title it is in many ways a new book.

It is also a tribute to the enormous contribution John Mazey (1926 - 2001) made to Tongariro National Park, the Park Ranger Service and, in particular, the role of interpretation in national park management whilst chief ranger at Whakapapa from 1961 to 1974.

"Whakangarongaro he tangata;
Toitu he whenua."
Man passes but land endures.

In memory of my sister, Marie Williams, and her fiancé, Derek White, who died in a helicopter crash on Mount Ruapehu — December 1982.

FOREWORD

My earliest recollection of the Tongariro National Park volcanoes was their sharp white outline seen from the Auckland-Wellington Express on a bright moonlight night in the late 1930s. Highlights since my appointment as Government Volcanologist in November 1945 included my first climb to the summit of Ruapehu that same month accompanied by Alan Beck and Horace Fyfe. An eruption that had begun in March was still in the explosive stage and each blast shook our narrow perch on Paretetaitonga. By early 1946 the volcano was quiet and with Alan Beck and Don Gregg I climbed from the head of the Whangaehu Valley and entered the crater over the south rim. At our feet lay the wide basin of the old crater within which was a central pit we estimated to be 300m deep. At the bottom was a small lake from which grey muddy water was continuously geysering. Only a year before, this had been the site of the well known Crater Lake — displaced in the interim by an expanding pile of black lava, shattered in turn by violent explosions, spread as ash over a large part of the North Island — and now this great hole. Should we get out quickly? I remember a feeling of awe and fascination at the apparent serenity of the scene.

Around active volcanoes tragedy can go hand in hand with excitement, generating a measure of respect for these uncontrollable forces. I inspected the devastation at Tangiwai on 26 December 1953. One hundred and fifty-one people had lost their lives two nights before when a large flood from Ruapehu's summit lake swept down the Whangaehu River demolishing the railway bridge and carrying away part of the Wellington-Auckland Express train. The flood, or lahar, was an aftermath of the 1945 eruption. Crater Lake had filled again and a barrier formed by the eruption had collapsed releasing the flood.

Considerable advance has been made during the last decade in the study of volcanic processes and this book with its abundant illustrations provides an up to date summary of this information. It describes the birth, growth and agents of change of the many landscape components and rocks of the volcanic terrain. On a broader scale it relates the geological setting of Tongariro National Park to New Zealand, the Pacific and the World. The volcanoes with their mantle of ice and snow have become the winter playground of the North Island. Their special and varied topography together with the retreat of glacial ice, particularly in the last 30 years, has focused increasing attention on the Park as a tramping and sightseeing area of unique interest. This book becomes a field guide to the visitor over the craters, lava flows, tephra layers and glaciated valleys that are so well exposed in the Park. The writer has compiled a book of wide interest that is a major contribution to New Zealand national park literature.

Jim Healy

A flat-topped cloud forms above Mount Ruapehu following a steam eruption from the Crater Lake in April 1982.

To the south, beyond the gently steaming summit crater of Ngauruhoe, lies the bulky form of Ruapehu — the largest volcano of the Tongariro Centre.

CONTENTS

"the Maori of New Zealand has his sacred Tongariro, which it is my hope, may long remain in its wild and chaotic grandeur."
Roderick Gray, 1890

A print of the southern shore of Lake Taupo and the volcanoes of Tongariro National Park in 1890. Artist unknown.

INTRODUCTION

Volcanoes have long been regarded with superstition and fear by man who has used myth and legend to explain why the land erupts in fire and molten rock. Even the word volcano is derived from legend. It was once believed that Vulcan, the Roman god of fire, worked in a giant blacksmith's forge beneath Mount Etna keeping the fires of this volcano burning.

The early Maori's conception of volcanology was also based on myths about gods and demons. It is alleged that they brought volcanic fire to New Zealand from Hawaiki, the Pacific homeland of the Maori.

Following the discovery of mountains in the central North Island, the priest Ngatoro-i-rangi decided to climb one of these peaks and claim the surrounding land for his tribe, the Ngati Tuwharetoa. On reaching the summit in a snowstorm, Ngatoro-i-rangi was almost frozen and his strength failing. In desperation he called to his sisters in Hawaiki to send him fire —

"Ka riro au i te tonga! Haria mai he ahi moku!" ("I am borne away by the cold south wind! Send fire to warm me").

Ngatoro-i-rangi's priestess sisters heard his request. They sent fire demons by way of an underground passage to White Island, Rotorua, Taupo and finally to the mountain top where he stood. Fire burst forth and Ngatoro i rangi was saved. An unfortunate slave, Auruhoe, was thrown into the blazing crater to appease Ruaimoko, the volcano god.

Thereafter, the volcano was called Auruhoe but more recently became known by its present name of Ngauruhoe.

The origin of the word Tongariro is contained in Ngatoro-i-rangi's prayer for fire — "tonga" (south wind) and "riro" (carried away). The name Tongariro was often applied to all three volcanoes (Tongariro, Ngauruhoe and Ruapehu) by the Ngati Tuwharetoa people.

Science has succeeded in stripping away the myths from the volcanoes. Today, we seek explanations based on an understanding of the geological processes that cause volcanic activity. We now know something about why volcanoes occur where they do and how they work. However, the awesome power and unpredictable nature of volcanoes ensures these "burning mountains" will retain much of their mystery and fascination.

THE MOBILE CRUST

A GLOBAL PERSPECTIVE

Man's view of planet Earth has changed a lot since the time he thought the world was flat. Our knowledge of the Earth's interior has increased through the study of earthquakes (seismology), and in the last 20 years a sophisticated theory called plate tectonics has been developed and used to explain the Earth's fundamental geological features.

The basis of this theory is that the crust of the Earth is made up of slabs of various sizes, rather like a giant jigsaw puzzle. These slabs, or plates, up to 60km thick, are more or less rigid and slowly move over the upper mantle of the planet. Over millions of years the great crustal plates have moved relative to each other, splitting apart in some places, colliding in others, and sliding past one another elsewhere.

Most volcanoes occur along these plate boundaries. Earthquakes tend to take place in the same zones. Together, volcanic and seismic activity usually define the edges of the Earth's crustal plates.

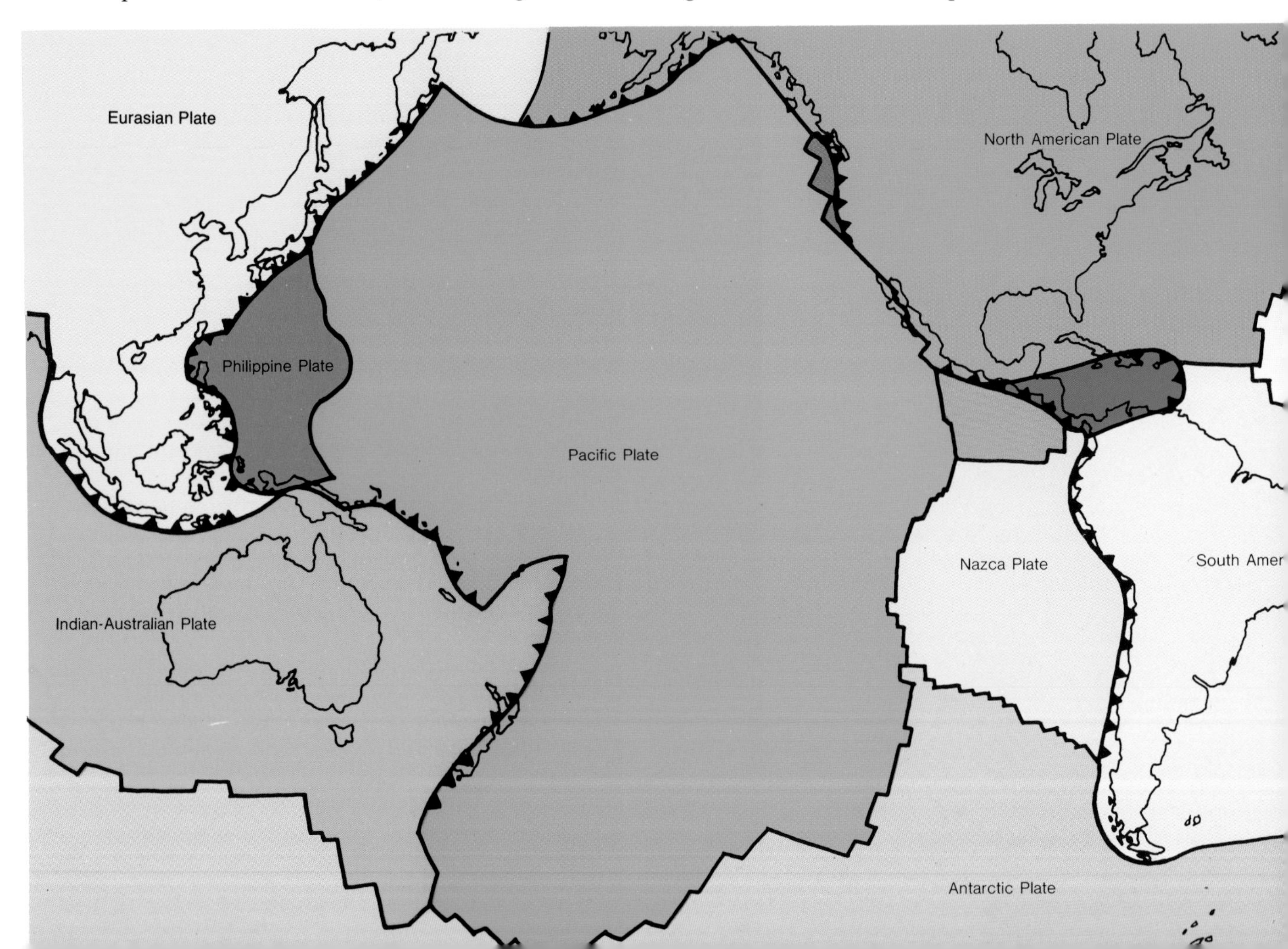

Volcanoes and earthquakes have a common origin in the movement of crustal plates. By plotting the patterns of earthquakes and volcanoes worldwide, scientists have been able to pinpoint the edges of at least seven large plates and a number of smaller plates.

The existence of the circum-Pacific chain of volcanoes, called the "Ring of Fire", can now be explained by plate tectonics theory. These volcanoes occur along the margins of the enormous Pacific Plate and the adjoining plates.

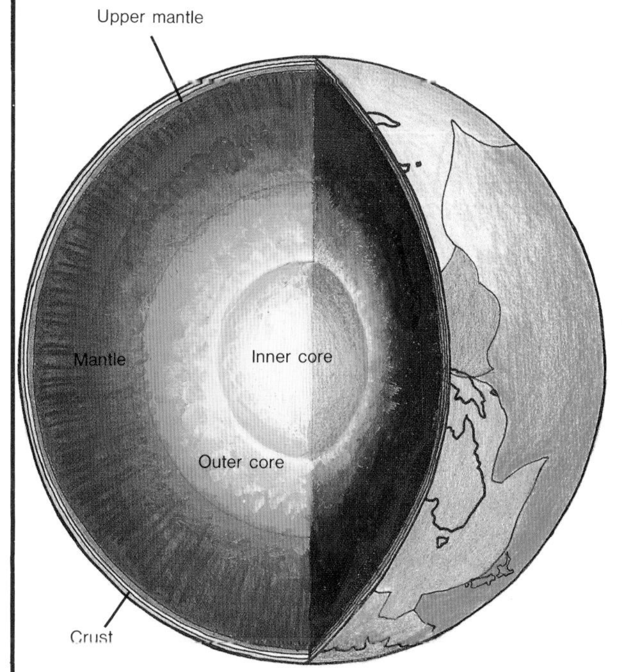

Seismic data have revealed that the Earth is composed of a number of concentric layers that have different physical and chemical properties. This cut-away diagram shows the main interior subdivisions of the Earth.

The crust of the Earth ranges from 12km to 60km in thickness and its rocks are cool and rigid. Beneath this thin outer "skin" lies the solid, hot rock of the mantle, a layer of about 2800km which comprises the bulk of the Earth. Beneath the mantle lies the Earth's core. It is composed of a fluid outer core (about 2000km thick) and a solid inner core (about 1400km thick).

Volcanoes have their origin in the crust and in the partially molten zone of the upper mantle which separates the crust and the mantle.

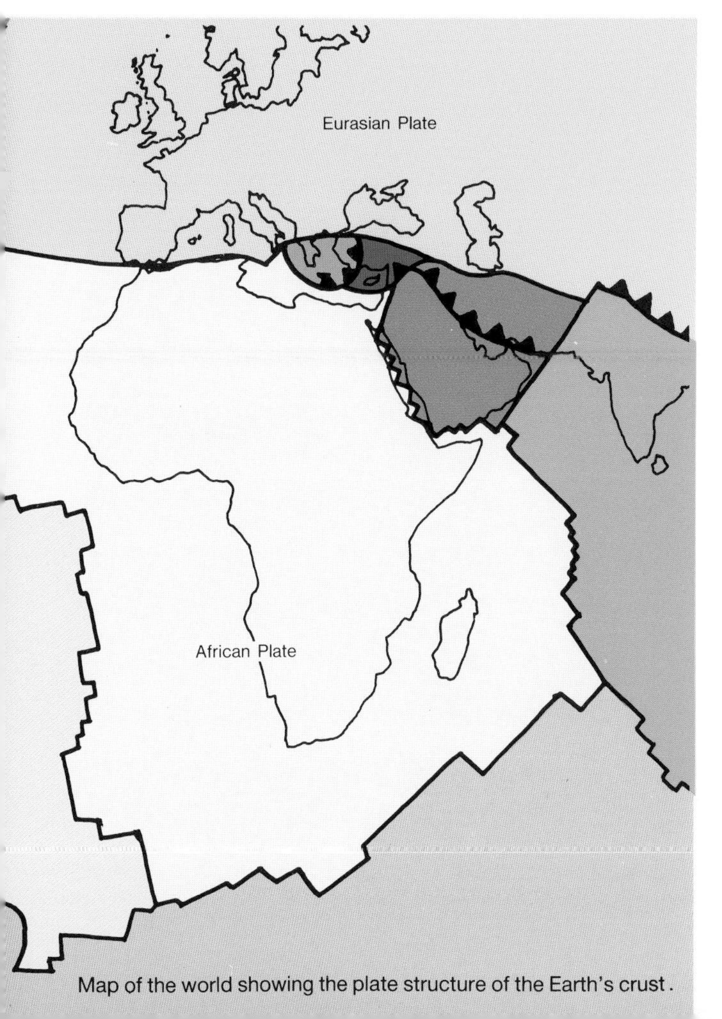

Map of the world showing the plate structure of the Earth's crust.

NEW ZEALAND ON THE EDGE

The present geology of New Zealand has largely been determined by its position on the edge of two crustal plates. The boundary between the Indian-Australian Plate and the Pacific Plate bisects the country. When two plates collide, one is forced beneath the other in a process known as subduction. Off shore, deep ocean trenches mark the subduction zones where the plates converge.

It is a complex situation. To the north, the Pacific Plate is descending beneath the Indian-Australian Plate, along the Tonga-Kermadec Trench. To the south of New Zealand the situation seems to be reversed. Here, the Pacific Plate is overriding the Indian-Australian Plate along the Puysegur Trench. The Alpine Fault, a prominent fault running for about 400km through the South Island, appears to connect these two subduction systems.

The Alpine Fault and, in fact, most of the country's major faults trend in a north-east to south-west direction parallel to the plate margins. Sediments (mud, silt and sand) that accumulated in the trenches along the edge of the plates have been compressed and uplifted along these faults to form many of the mountain ranges of New Zealand. The ranges are also aligned north-east to south-west.

New Zealand's twisted and buckled shape has resulted from a very long process of construction and deformation along the plate boundary.

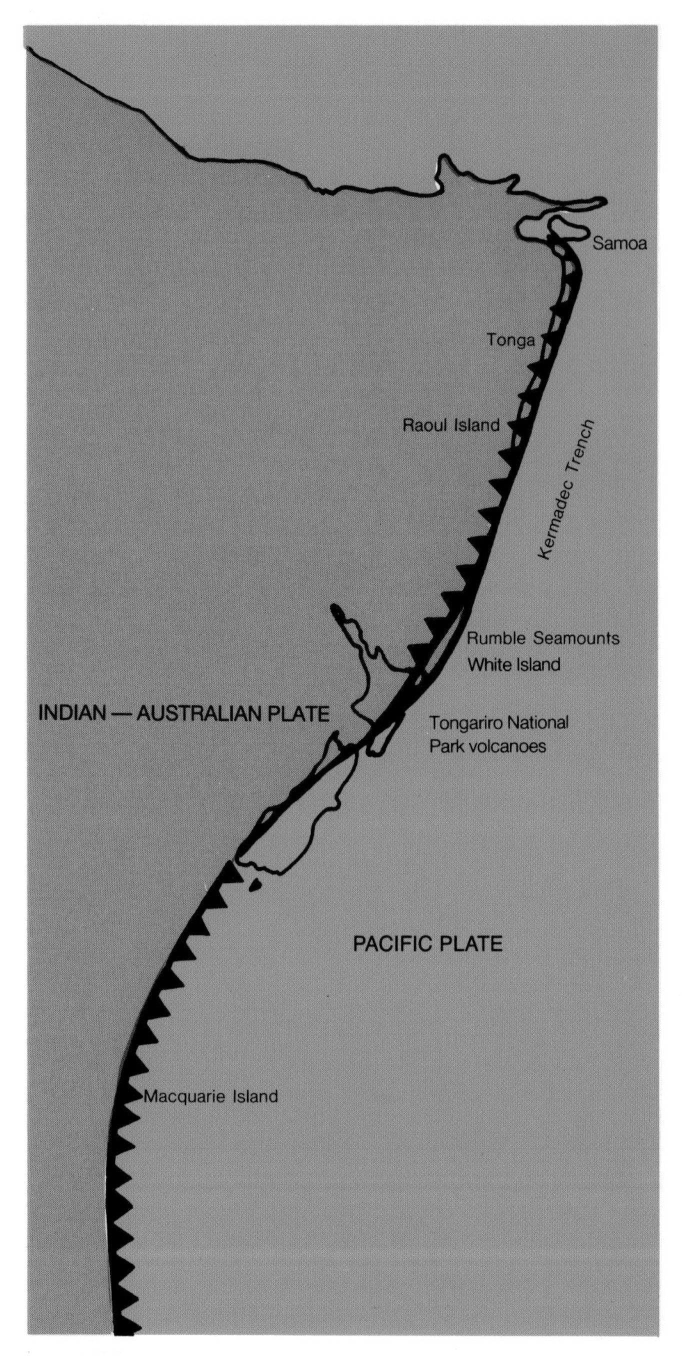

The volcanoes of the central North Island of New Zealand are part of a discontinuous chain of volcanoes extending north-east into the Pacific for about 2500km. The volcanoes of Tongariro National Park lie at the southern end of this chain which passes offshore to White Island and on to the active underwater volcanoes called the Rumble Seamounts. The line of volcanoes is continued along the island arcs of Kermadec and Tonga. Clearly, these volcanoes have a common origin. They are found along the edge of the overriding Indian-Australian Plate, parallel to the Tonga-Kermadec subduction zone. The Hikurangi Trench, off the East Coast of New Zealand, is a southern extension of this system.

The islands of Tonga also lie along the boundary between the Indian-Australian and Pacific plates. The volcanic cone of Tafahi (left) is an island in the Niuatoputapu group, northern Tonga. A deep ocean valley called the Tonga Trench runs parallel to the islands and reaches depths of 10km.

The volcanic trend is continued along the island chain of the Kermadec Group. The most recent eruption took place from Green Lake Crater on Raoul Island, the largest island in the group, in 1964.

White Island is an active andesite volcano situated in the Bay of Plenty 50km north of Whakatane. The central floor of White Island's crater is just above sea level.

Three active hydrothermal sites, known as the Calypso vents, have been identified about 15km south-west of White Island. They lie on the sea floor, at depths of more than 170m, and vent high temperature (200°C) gas-rich fluids from deep within the Earth.

The Hikurangi Trench marks the position where, in the plate tectonic process, the Pacific Plate dives beneath the Indian-Australian Plate. As if part of a slowly-moving (45mm a year) giant conveyor belt the descending plate of wet crust pushes deep into the Earth.

The Pacific Plate descends into a region of intense heat and pressure in the slowly convecting upper mantle. Molten rock known as magma is created at a depth of about 100km beneath the central North Island of New Zealand. The melting is caused by volatile material such as water in the descending crust reacting with the surrounding rock material. The fluid magma that results is less dense than the surrounding rock and it rises upwards. In places, the magma reaches the surface to erupt. Over time, the products of these eruptions have created the volcanoes of the central North Island.

Geothermal fields associated with volcanic and fault activity attest to high heat flow throughout the region. The Taupo Volcanic Zone is a region of crustal extension and is widening asymmetrically at a rate of about 7-18mm each year.

An artist's impression of the subduction zone beneath the North Island of New Zealand showing the relationship of the central volcanic region to the dipping Pacific Plate.

THE VOLCANIC ZONE

Most of New Zealand's volcanoes and geothermal features are found in the Taupo Volcanic Zone in the central North Island. It is a narrow belt of activity, only 20-40km wide, but spans a distance of 240km from Ohakune to White Island. This north-east trending zone is tectonically active, that is, it is characterised by earthquakes and active faults.

Volcanoes have been active in the Taupo Volcanic Zone for about one million years. During this time, eruptions have occurred from five major centres — Okataina, Rotorua, Maroa, Taupo and Tongariro. The volcanoes are a surface manifestation of the subduction process that is occurring beneath the North Island. Seismic evidence suggests that the Pacific Plate begins to descend steeply beneath the Taupo Volcanic Zone. Heat and pressure intensify as the plate moves further into the Earth, and rocks begin to melt.

Molten magma rising from different depths (between 75km and 150km) below the zone varies in composition. This has resulted in the diversity of eruption styles and landforms that are a feature of the Taupo Volcanic Zone.

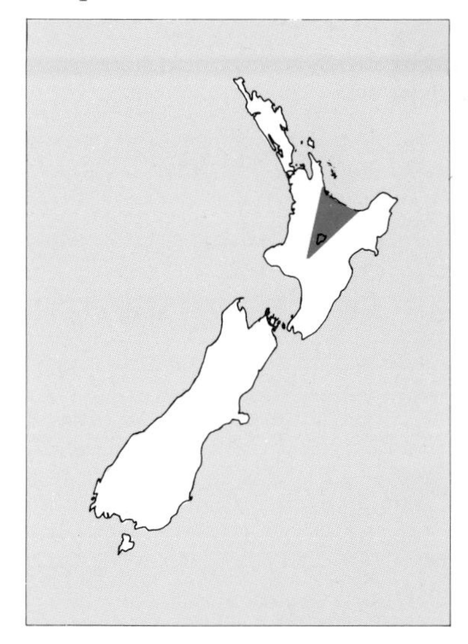

Map of New Zealand showing the location of the Taupo Volcanic Zone.

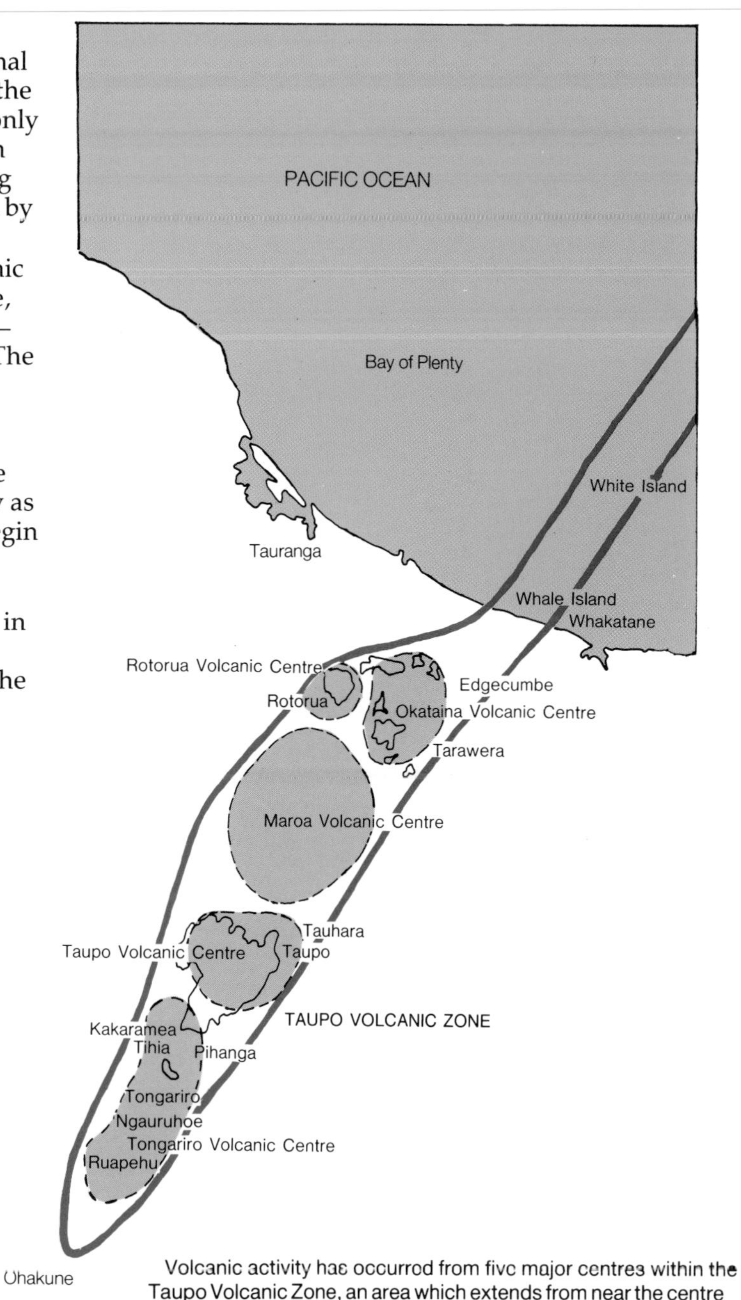

Volcanic activity has occurred from five major centres within the Taupo Volcanic Zone, an area which extends from near the centre of the North Island to the north-east into the Bay of Plenty.

VOLCANOES

MAGMA — THE RAW MATERIAL

Magma is the raw material of volcanoes. It is made up of varying proportions of (hot) liquid rocks, volcanic gases and solid particles. It is this diversity of chemical composition, between different magmas, which leads to such wide variations in the size and style of volcanic eruptions.

Silicon dioxide (SiO_2), commonly called silica, is an important ingredient of magma. Magmas and the volcanic rocks they form are divided into three broad groups according to the abundance of silica in them.

The amount of silica in a magma influences its viscosity. This term is used to describe the way in which a liquid (in this case magma) flows. A silica-rich magma will not flow easily and is said to have high viscosity. In contrast, a silica-poor magma flows readily and has low viscosity. This property becomes crucial as magma nears the surface because viscosity affects the way volcanic gases trapped in the molten mix escape.

Volcanic rocks containing a high proportion of silica are known as acidic rocks. Intermediate rocks contain slightly less silica, and basic rocks, the least. Three principal volcanic rock types correspond with these groups — rhyolite, andesite and basalt.

Rhyolites typically contain the minerals: quartz, feldspar and biotite. They are rich in silica (SiO_2), soda (Na_2O) and potash (K_2O). The glass formed by rapid cooling of liquid magma is also rich in these oxides. As quartz, feldspar and glass are all light in colour, rhyolite lava is generally a light-coloured rock.

More than 67% silica
ACIDIC

Andesite is intermediate between rhyolite and basalt and, as one might expect, ranges in colour from light grey to black, depending on mineralogy and composition.

53-67% Silica
INTERMEDIATE

Basalts contain pyroxene, olivine and basaltic glass which are rich in magnesium and iron oxides ($MgO + FeO$). These minerals and glass are dark in colour, so basalt is generally black. If the iron is strongly oxidised and forms ferric oxide (Fe_2O_3), it turns red. Basalt scoria is commonly red in colour.

Less than 53% Silica
BASIC

VOLCANIC GASES — THE DRIVING FORCE BEHIND ERUPTIONS

During an eruption, magma moves in response to immense pressures from within the volcano. As the molten rock nears the surface, pressures are lowered, and water and other gases boil off. The gases may then escape explosively or gently, depending on their abundance and the composition of the magma they are contained in. The potential violence of an eruption increases with increasing gas content of the magma. Thus, the way the gases escape largely determines the type of eruption.

From the extremely fluid (low viscosity) and hot basic magmas, volcanic gases escape easily and without pressure. Basalt eruptions are typically "quiet".

Gases have more difficulty escaping from the more viscous intermediate magmas. Consequently, andesite eruptions tend to be relatively violent.

Gases escape with difficulty from acidic and cooler rhyolite magmas which are characteristically highly viscous. These magmas are less able to release the pressure of the contained gases, and extremely violent explosive eruptions can result.

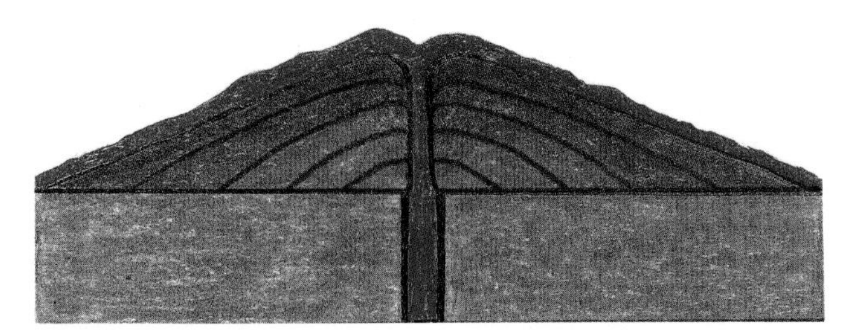

A quiet eruption

A moderate eruption

An explosive eruption

WHAT COMES OUT OF VOLCANOES?

A volcano does not always erupt in the same manner or emit material in the same form. What comes out of a volcano depends on the chemical composition and physical characteristics of the magma feeding an eruption. This influences the nature of the eruption and the final volcanic product.

Lava eruption, Hawaii

Lava

Lava is the product of non-explosive eruptions. It is the collective term for all molten material, of any composition, which flows from a crater. It is the same material as magma, but without the gas content. (The volcanic gases which were dissolved in the magma have been liberated to the air.) As it cools, lava loses its ability to flow and solidifies to form volcanic rocks.

Molten lava, like other liquids, flows downhill under the influence of gravity. Rhyolite lava moves with great difficulty, forming short, stubby flows. Andesite flows more easily than rhyolite lava and forms thick, tongue-like flows. Highly fluid basalt lava spreads out to form thin and extensive sheets.

Pyroclastic Rocks

When volcanoes erupt explosively, the force shatters magma and rocks lining or blocking the vent. This is the origin of the fragmented material ejected through the air from volcanoes. The term pyroclastic (meaning "fire-broken") is used to describe this material. Eruptions of pyroclastics generally occur from volcanoes whose lava is relatively viscous. Fragments, shot into the air, form an eruption column above the vent before falling back to the ground to blanket the countryside with "airfall" deposits. Sometimes an eruption column collapses and pyroclastic material "flows" over the surface of the ground. Both airfall and flow deposits form layers, called tephra layers. Tephra is the Greek word for ash.

Classification of Pyroclastic Rocks

Fragments of solid lava ejected from volcanoes are called pyroclastic rocks or tephra. These fragments are classified according to size and shape:

ash —	less than 2mm diameter
lapilli —	between 2mm and 64mm diameter
blocks —	angular fragments greater than 64mm which were ejected as a solid
bombs —	round fragments greater than 64mm which were ejected in a wholly or partly molten condition.

Ash eruption, Mount Ngauruhoe

Pumice is produced when rapidly expanding gases cause rising magma to froth. This foamy material is erupted explosively and quickly cools to become pumice. The expansion and escape of gases produce the small spherical cavities that make pumice light enough to float. Pumice is usually erupted from rhyolite volcanoes.

Scoria is derived from material that is rapidly cooled while travelling through the air. It is honeycombed with small holes which are formed by gas bubbles in the still-liquid fragments. Scoria ranges in colour from grey to black and red when it is ejected, and is usually found on andesite and basalt volcanoes.

THE SHAPE OF VOLCANOES

"I think it is possible to make a generalisation: the more like a volcano your pile looks, the smaller the eruptions . . . it is the harmless-looking things like Taupo, which don't seem like volcanoes at all, that have given rise to some of the most violent eruptions on our planet."

George Walker, 1978

The shape of a volcano is determined primarily by the composition of the magma feeding it, the style of past eruptions, and the material ejected during those eruptions.

Andesite magma, for example, may be erupted quietly or explosively, sometimes producing lava, sometimes pyroclastics. When eruptions of this kind occur from a central vent a steep, cone-shaped volcano often develops.

Andesite eruptions typically build volcanic landforms called composite cones, or strato-volcanoes. These structures consist of layers of pyroclastic material, irregularly alternating with lava flows.

Composite cones are not always symmetrical. Eruptions can take place from vents on their flanks and result in the construction of "parasitic cones". These vents are fed by magma which probably stems from the central conduit of the volcano. Magma forces its way through the walls of the cone, cutting across beds of lava and tephra. This type of activity leads to the formation of geologic features called dikes (or dykes).

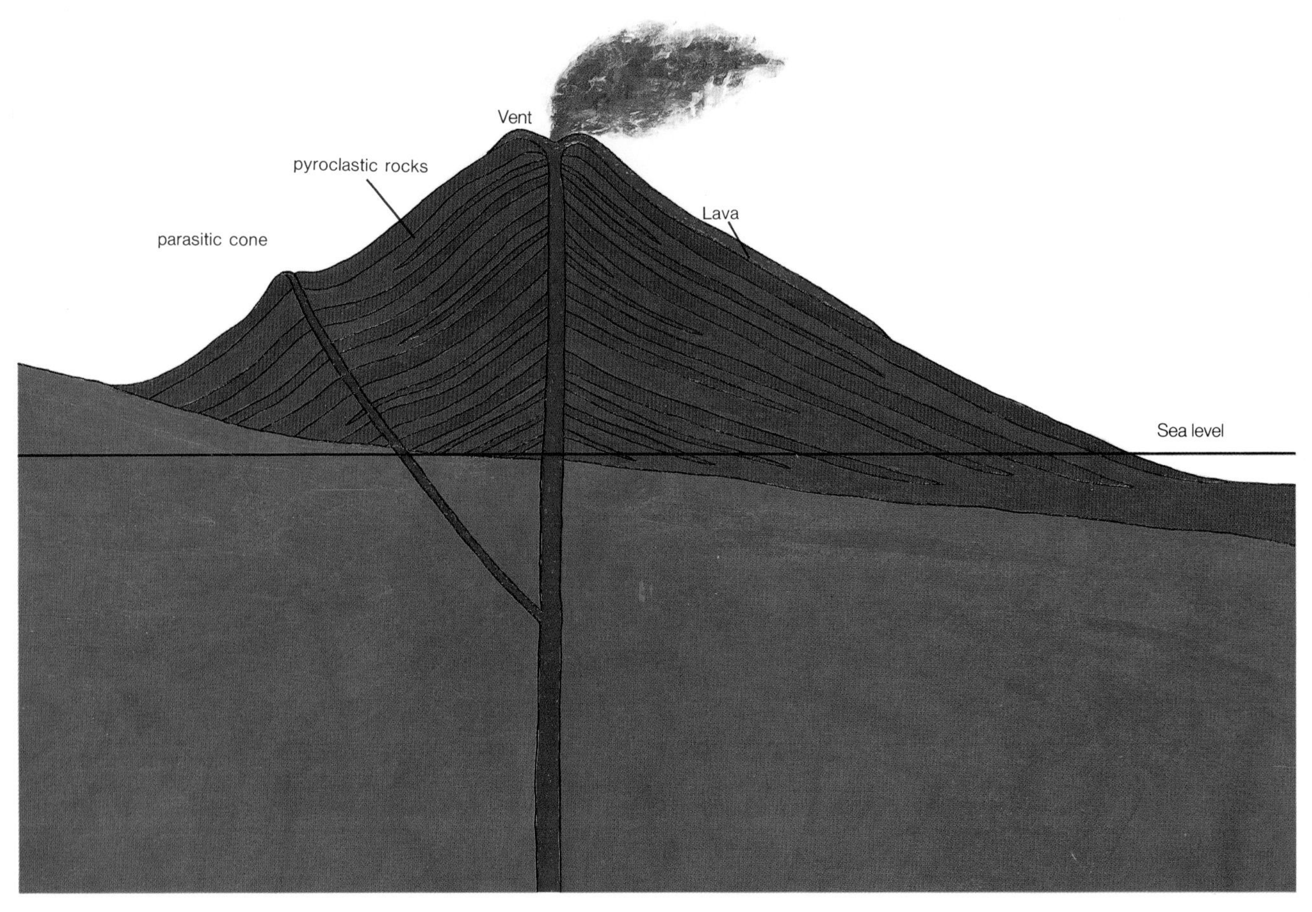

Andesite Cones

Andesite volcanoes erupt both solid pyroclastic material and liquid lava. Relatively moderate eruptions pile this material around the active vent. Eruptions occurring intermittently over a long period from a central vent will slowly construct a steep-sided conical landform. Mount Ngauruhoe, one of the volcanoes of the Tongariro Centre, is a classic example of a composite andesite cone.

Mount Ngauruhoe.

Lava Domes

Viscous silica-rich magma, rising slowly in the vent, sometimes causes the ground surface to bulge up into a dome of internal growth. As lava continues to be squeezed up from beneath, the bulbous form continues to quietly expand.

More than 150 lava domes are found in the Taupo Volcanic Zone. Many of these are rhyolite domes, including the rhyolitic Tarawera lava domes of the Okataina Centre. Domes of dacite lava (formed from magma containing less silica than rhyolite but more than andesite) are also common in the zone.

Mount Tarawera.

Calderas

The most violent eruptions of all occur from gas-rich rhyolite magma. Gigantic explosions blow out vast amounts of pyroclastic material over a wide area. This can lead to the collapse of the roof of the volcano above the now-empty magma chamber. The resulting landform is a large basin-like depression called a caldera, which may later fill with water. Lake Rotorua fills a portion of a caldera that is about 15km wide.

An outline of the rim of the Rotorua caldera has been superimposed onto this satellite photograph. Calderas are also found in the Taupo, Maroa and Okataina Centres of the Taupo Volcanic Zone.

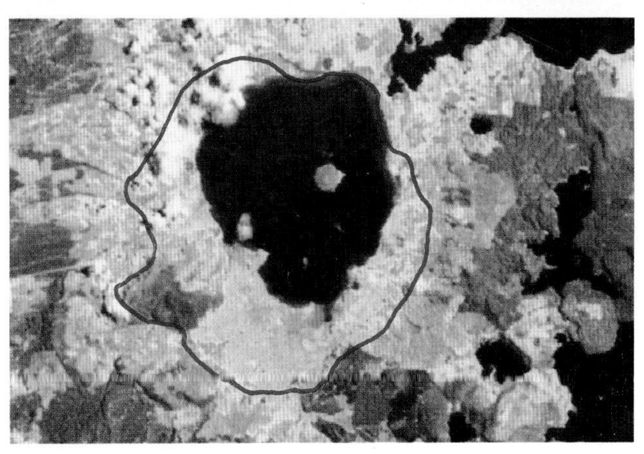

Rotorua caldera.

TONGARIRO VOLCANIC CENTRE

THE GEOLOGICAL TIME FRAME

The volcanoes of the Tongariro Centre lie at the southern end of the Taupo Volcanic Zone. Most of this area comprises the Tongariro National Park. The volcanoes stretch from the southern shores of Lake Taupo to Ohakune, a distance of 55km. In terms of geological time, which on Earth extends back to the origins of our planet about 4600 million years ago, these volcanoes are extremely young.

The oldest dated lava flows from the Tongariro Centre are 260 000 years old, but the first eruptions probably began about one million years ago in the middle of the geological period called the Quaternary. However, andesite pebbles found near Wanganui indicate that volcanism was going on in the region about 1.7 million years ago. During this time the area was above sea level.

Before this, about 5-15 million years ago, the area formed part of a shallow marine basin in which sediments accumulated. Sediments containing sea shells from this Tertiary Period can be found today as outcrops in the south of the Park.

The oldest rocks of the area developed earlier still, during a period called the Mesozoic. Sediments known as the Torlesse Greywackes were deposited in the sea about 150 million years ago. The Kaimanawa Range to the east of the Park and Taurewa Range to the west are composed of rocks of this type. These ancient sedimentary rocks are also believed to form the "basement" on which the volcanoes developed.

The closely grouped volcanoes of the Tongariro Volcanic Centre lie to the south of Lake Taupo.

"Volcanoes dominate Tongariro National Park and give the area its distinctive character. Their eruptions have blanketed the surface with ash, and lava has flowed on to the surrounding plains. They form a striking southern terminus to the Taupo Zone of Quaternary volcanism."

Donald Gregg, 1960

HOW OLD IS THIS VOLCANIC LANDSCAPE?

The volcanic rocks and landforms exposed at the surface of the Park are not all the same age. They span a period of about 300 000 years. By dating the rocks and tephra layers making up this landscape, a sequence of events can be established and a picture of the evolution of the Tongariro volcanoes compiled.

Many of the features of the Park's landscape can be dated with respect to two of the most significant events to affect this area — the Ice Age and the Taupo eruption.

USING THE ICE AGE TO DATE THE LANDSCAPE

The last glaciation, commonly called the "Ice Age", maintained its icy grip on much of the world for close to 100 000 years. During this period the South Island of New Zealand was extensively glaciated, while the volcanoes of the Tongariro Centre were among the few peaks in the North Island to support glaciers. Perennial snow covered the mountains down to about 1500m, 1000m lower than the present snowline. Glaciers on Tongariro and Ruapehu expanded throughout the cold periods of the Ice Age, extending at their maximum to about 1200m, the elevation of Whakapapa Village.

The erosive power of the glaciers altered the shape of the existing volcanoes in characteristic ways. As the ice streams moved, they also disturbed layers of volcanic ash in their path. The oldest tephra layer in the Park not eroded by glacial action is about 14 000 years old. Therefore, it is believed the Ice Age was coming to an end in the area at this time.

Since then, volcanic activity has created new landforms untouched by the glaciation. Thus, the physical features of the Park can be divided into two broad categories: landforms shaped by glacial erosion which are at least as old as 14 000 years; and landforms that are younger than the last glaciation. The age of post-glacial landforms can be further narrowed down by establishing whether they are younger or older than the Taupo eruption.

The Wahianoa Valley on the south-eastern slopes of Mount Ruapehu was shaped by ice. This broad valley still bears the imprint of the last glaciation.

TAUPO CALDERA

The caldera volcano of Lake Taupo has a complex history dating back about 300 000 years. The huge Oruanui eruption of 26 000 years ago is the largest eruption known in the history of the volcano, and is thought to be largely responsible for the shape of the present-day lake. The Oruanui event has been followed by 28 eruptions which have occurred during the past 21 000 years.

The volcanic record, found in road-side cuttings in the region, reveal that these numerous young eruptions occurred from different magma systems and from vents now located beneath the modern lake.

Some eruptions have been relatively small, but the most recent, the Taupo eruption of 1800 years ago was one of the most powerful in the world in the past 5000 years. Pumice was ejected to form a column that at times reached heights of more than 50km. The central North Island was showered with rhyolitic airfall deposits of white and pale yellow lapilli and ash, to depths of up to 5m.

The giant eruption column eventually became unstable and collapsed to generate a pyroclastic flow of fragmented volcanic material. This flow spread out from the vent at speeds of up to 1000km per hour. Within ten minutes the pyroclastic flow had levelled forests and buried vegetation within a radius of 80km of Taupo, including much of Tongariro National Park. Pumice, ash and other volcanic debris filled valleys and depressions and cooled to form a layer, locally up to 30m thick in the park, of rock known as ignimbrite.

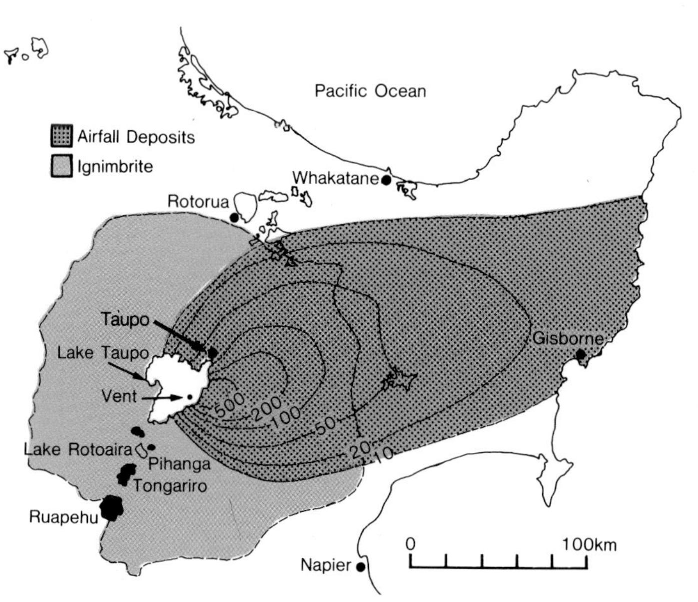

A map of the eastern central North Island showing the total thickness (in centimetres) of airfall deposits from the Taupo eruption. The area covered by ignimbrite from Taupo is also shown.

The distinctive deposits from the 1800-year-old Taupo pumice eruption, the most recent eruption from Taupo, provide an important dating layer throughout the central North Island. Carbonised logs and pieces of wood buried within the deposit are evidence that these hot pyroclastic flows burned all vegetation in their path.

The volumes of tephra erupted from Taupo in about A.D.186 and from Mount Ngauruhoe in 1975 are compared with some well known eruptions from other volcanoes. This illustrates that the 1975 eruptions of Ngauruhoe, although locally spectacular, were very small on a world scale.

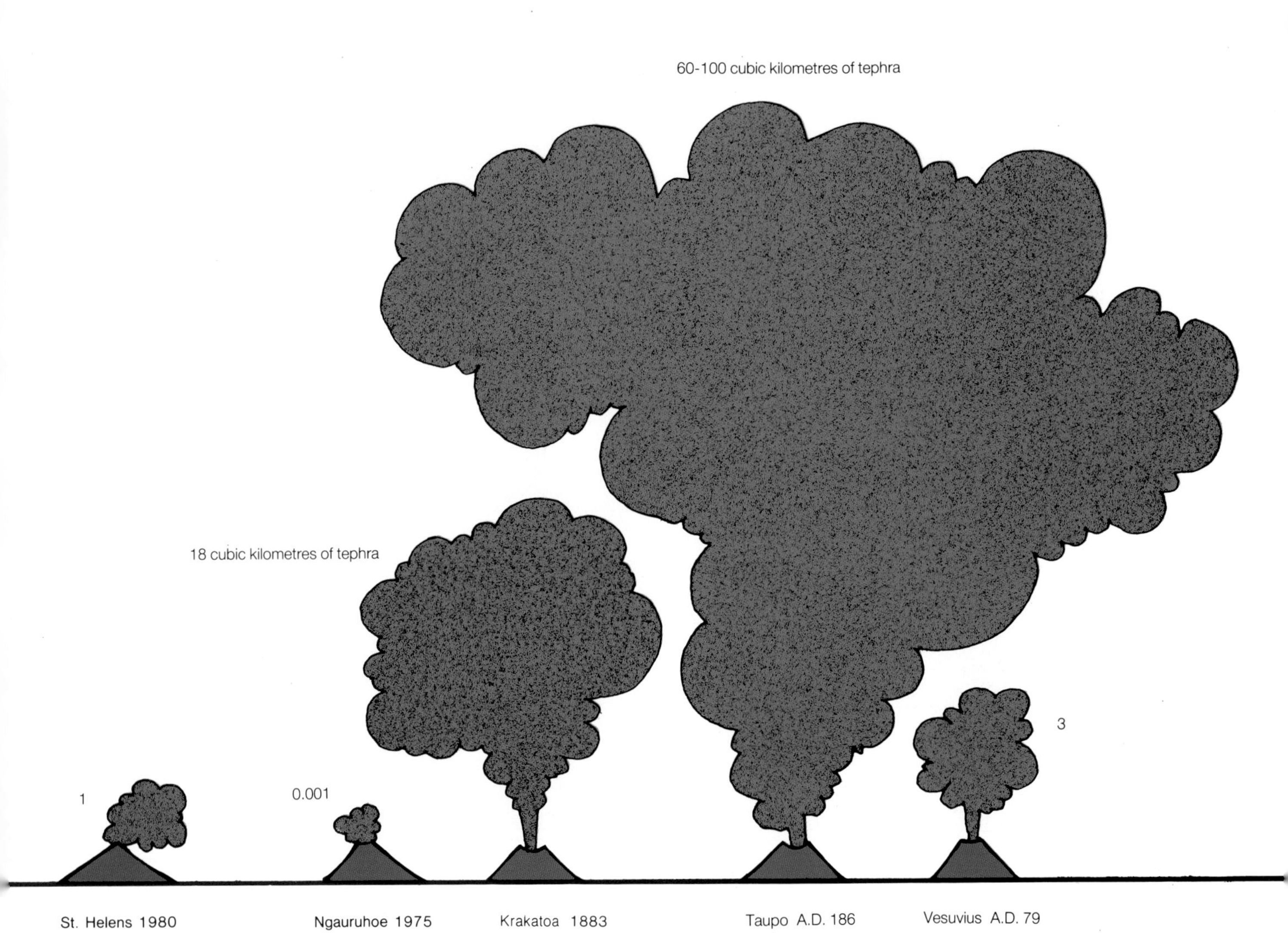

LAYERS OF VOLCANIC HISTORY

Formations	Thickness	Description	Source
Ngauruhoe	0.22m	Very dark brown fine ash	Ruapehu (Ngauruhoe, Tongariro)
Taupo Pumice (AD 186)	0.6m	White pumice, lapilli and ash	Taupo
Mangatawai Tephra	0.21m	Dark brown and dark grey ash with beech leaves	Ngauruhoe
Numerous andesite tephra units from Tongariro, Tama Lakes area and Ruapehu			
Rotoaira Lapilli	2.0m	Yellow-red & yellow-brown ash & lapilli units. Basal sands with olivine andesite cobbles from Pukeonake	Te Maari Craters
Oruanui or Wairakei	3.6m	Grey-brown ash with accretionary lapilli	Taupo

Layer upon layer of tephra, representing countless eruptions from the Tongariro Centre and other volcanoes in the Taupo Volcanic Zone, form a deep ground cover in parts of the Park. Each layer is a record of a past eruption and geologists study these deposits to find out the frequency, extent, magnitude and source of the eruptions. Individual layers can be distinguished on the basis of colour, size of fragments, mineral content and the thickness of the deposit.

The age of a tephra layer and therefore the date of the eruption which led to its formation can be determined if peat, charcoal or wood are associated with the deposit. The ages of many tephra layers in the central North Island are now known with reasonable certainty and these provide valuable time markers. Two rhyolitic tephras, the 1800-year-old Taupo pumice deposit and the Oruanui Formation (a 20,000-year-old tephra from the Maroa Volcanic Centre), have proved to be very useful marker beds for dating the landforms of the Park.

THE VOLCANOES OF THE PARK

The volcanoes of Tongariro National Park are predominantly andesite in composition. Andesite magma has come to the surface along fractures or zones of weakness associated with the subduction of the Pacific Plate. The word andesite is derived from the Andes, the volcanic mountain chain of South America, which also formed parallel to a subduction zone.

The Park encompasses a landscape that has been the scene of intense volcanic activity. The signs are everywhere: in the craters, lava flows, explosion pits, tephra layers, scoria and composite cones. This book focuses on the volcanic landforms and other distinctive features of the area, working from north to south through the Park.

The volcanoes of the Tongariro Centre can be separated into two groups on the basis of location, activity and size.

Diagram showing the location of the two volcanic chains which together comprise Tongariro National Park

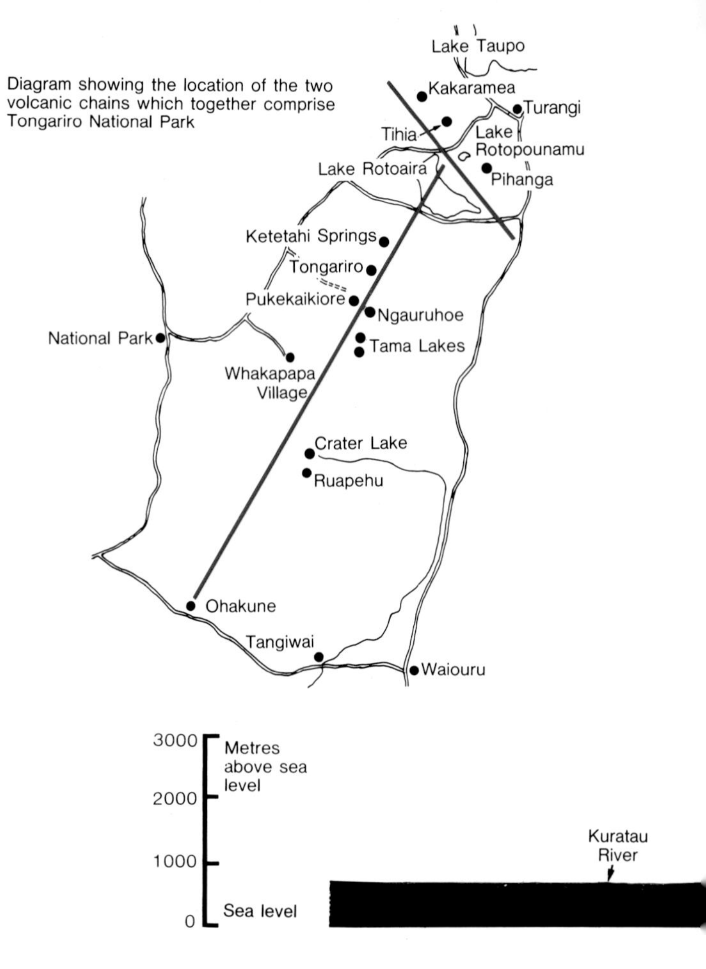

3000
Metres above sea level
2000
1000
Kuratau River
0
Sea level

A cross-section through the Southern Volcanoes

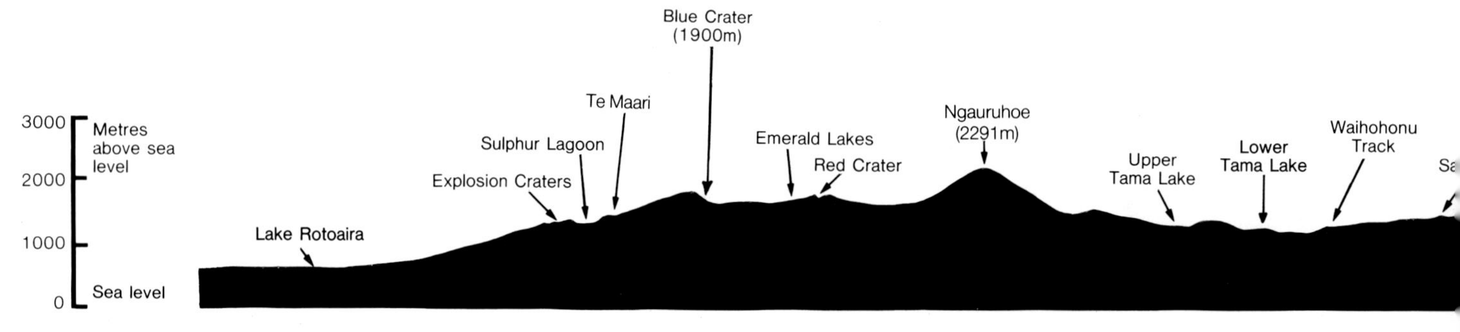

Tongariro

Two composite andesitic volcanoes, Kakaramea-Tihia and Pihanga and their associated vents, compose the northern volcanoes. This short chain (10km long) lying on the Park's northern boundary, erupted along a line trending north-west to south-east almost at right angles to the present regional pattern of volcanism. These extinct weather-beaten peaks are little more than 1000m in elevation, about half the height of the still active volcanoes to the south.

The younger, multiple vents of Tongariro and Ruapehu form the southern group of volcanoes. The vents of the southern chain are aligned north-east to south-west, corresponding to the present active trend of the Taupo Volcanic Zone and the trend of the Taupo Fault Belt.

Looking from the daisy-covered slopes of Mount Tihia across Rotoaira to Mount Tongariro.

A cross-section through the Northern Volcanoes

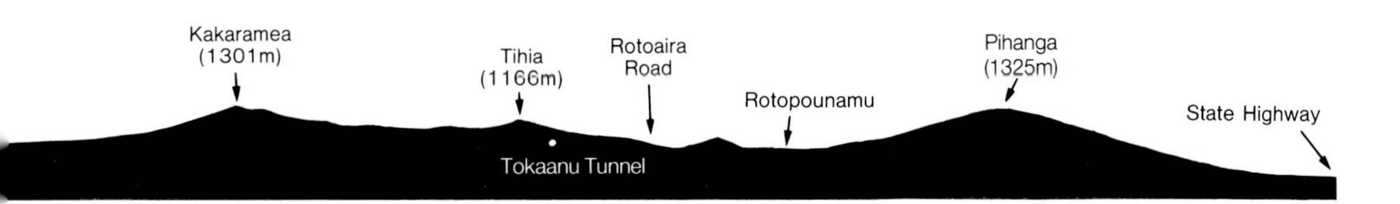

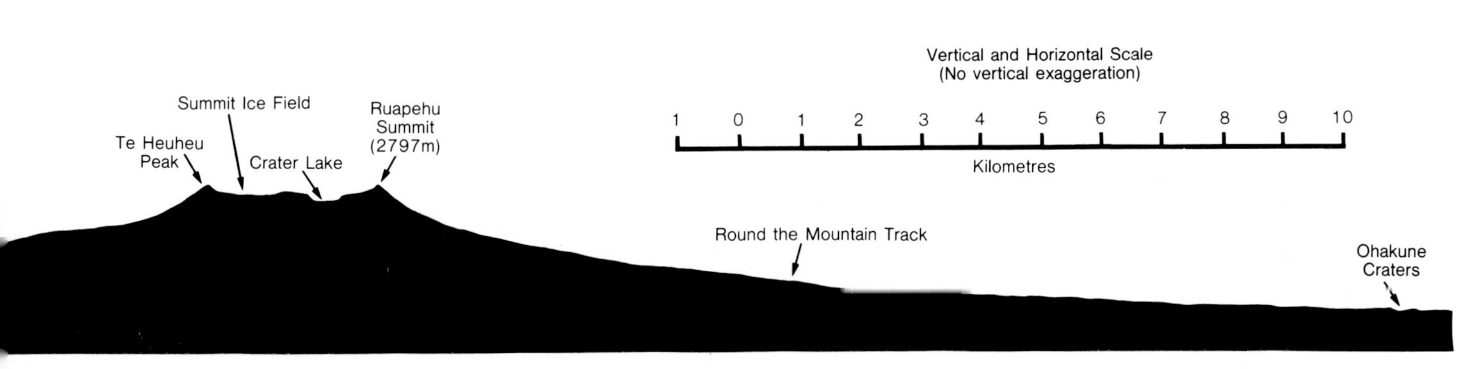

Ruapehu

THE NORTHERN VOLCANOES

KAKARAMEA-TIHIA MASSIF

The low ridge to the south of Lake Taupo is composed of a number of closely grouped volcanic vents which have merged together to form what geologists call the Kakaramea-Tihia massif. The 1300m andesite cone of Mount Kakaramea forms the summit of this old volcano and the lava dome of Mount Tihia lies 3.5km to the south-east. Several other domes, cones and craters occupy a broad area of swampy ground between the two peaks.

The eruptive history of this multiple volcano is not well known. Dates obtained from lava flows suggest acitivity took place here between 100 000 and 230 000 years ago. However, the youngest dated tephra deposit, from a vent near the eastern end of the massif, has a radiocarbon age of 40 300 years.

Faulting has extensively modified the shape of the Kakaramea-Tihia massif. Major north-east trending faults of the Taupo Fault Belt crisscross the area, forming clear fault traces.

Water and steam heated by magma beneath the mountain, reach the surface along a fault-scarp on its northern slopes. Although this area, called the Hipaua Thermal Area or "the steaming cliffs", is beyond the Park boundary there are strong geological and historical links with Tongariro National Park.

Hydrothermal activity on the hillside alters the volcanic rock to clay, weakening the face of the scarp. A number of mudflows have occurred in the area during geologically recent times. A large mudflow swept through the village of Te Rapa, east of the present village of Waihi, in 1846. Fifty-five people were killed including Te Heuheu Tukino II, the paramount chief of Ngati Tuwharetoa. (It was Te Heuheu's son who later gave the mountain peaks of Tongariro, Ngauruhoe and Ruapehu to the Crown to form the nucleus of this national park.) Another large mudflow occurred in the vicinity in 1910. Landslide deposits are also abundant on the north-east slopes of Tihia.

Numerous small craters and domes can be distinguished on the summit of Tihia. Mount Pihanga, beyond and to the east, continues the northern chain of Tongariro Centre volcanoes.

The steaming cliffs of the Hipaua Thermal Area on the northern slopes of the Kakaramea-Tihia complex. The village of Waihi is nestled at the base of the mountain alongside the prominent fault scarp.

"the volcanic forces below have by no means been as yet lulled to their final repose; for on the northern declivity and at the foot of the Kakaramea it steams and boils in more than a hundred places."

Ferdinand von Hochstetter, 1867

ROTOPOUNAMU GRABEN

Rotopounamu graben is the technical name for the down-faulted block which lies between Mount Tihia and the western flanks of Mount Pihanga.

The cones of Pukepoto (930m) and Pukehohoao (795m) in the north and the Onepoto Craters in the south of the graben are believed to be the youngest vents in the northern group of volcanoes. They last erupted more than 20 000 years ago.

It is not known if the basin in the middle of the graben, filled by Rotopounamu, is an explosion crater of similar age. It could be a much younger feature that has formed as the result of faulting. Several streams drain into the lake which has no visible outlet.

Unlike the older part of the northern chain, the cones and craters of the graben are aligned north-east to south-west. They line up with the present active vents of Tongariro and Ruapehu and it is possible that volcanism could occur in the Rotopounamu graben in the future.

Lake Rotopounamu is bounded by young cones and craters to the north and south and by Mount Pihanga on its eastern side. A 20 minute walk from the road across Te Ponanga Saddle, between Turangi and Lake Rotoaira, leads to this beautiful lake. Rotopounamu, the "greenstone lake", is the largest lake in the Park.

MOUNT PIHANGA

Pihanga (1325m) lies at the eastern end of the Park's northern belt of volcanic activity. This cone-shaped andesite volcano is younger than the neighbouring Kakaramea-Tihia complex, remained active for longer and is less dissected by erosion.

The prominent scar, high on the north side of Mount Pihanga, is not a crater as is often thought but is the result of erosion and landsliding. A mantle of Taupo pumice and a thick cover of brown andesite ash from Tongariro vents during the past 15 000 years make it difficult to locate the last active crater. Its rim may be exposed at the head of the slip. Lava flows dipping to the south suggest that the vent was sited on the south-east of the summit of Pihanga, immediately to the south of the slip.

In times long since past, there were many more mountains in this region. They were all males except for the lovely maiden, Pihanga. Each of the gods wished to marry Pihanga, but she favoured Tongariro and it was he who eventually won her after a fierce fight. The defeated mountain gods were forced to leave the area. Putauaki (Mount Edgecumbe) and Tauhara travelled to the north, towards the morning sunshine; and Taranaki (Mount Egmont) to the setting place of the sun — the sea coast in the west.

A scar has been gouged in the northern face of Pihanga by land slips and erosion.

Lake Rotoaira occupies a down-faulted depression at the south-west base of Pihanga. Although the lake itself is not part of Tongariro National Park it separates the northern and southern volcanoes of the Park. The small island of Motuopuhi is a volcanic dome of dacite lava that was extruded more than 20 000 years ago. The northern slopes of Mount Tongariro are reflected in the lake.

THE TONGARIRO COMPLEX

This huge and sprawling volcanic complex extends from the shores of Rotoaira to Tama Saddle beyond Ngauruhoe. Potassium-Argon dating reveals the massif has grown almost continuously during the last 275 000 years and shows periods of rapid cone growth. Vents of all ages share the same alignment, a north-east/south-west aligned vent corridor, 5km wide and 13km long.

Early Tongariro was built from the overlapping and diverse eruptive products of at least six cones between 275 000 and 65 000 years ago. The complex was then immersed in snow and ice during the Ice Age and glacial erosion deeply scarred the landscape. However, Tongariro's modern shape may be due in part to gravitational failure as well as to glacial erosion and subsequent volcanic activity. Deposits from a huge landslide (4-26m thick and 15km long) have been identified on the north-west Tongariro ring plain from a source above the current Mt Tongariro summit.

Over the last 15 000 years, eleven post-glacial vents grew on the eroded remnants of older Tongariro cones. Isopach maps and stratigraphic records reveal an intense period of volcanic activity about 10 000 years ago when multiple vents were active, either simultaneously or sequentially along a 10km line between Te Maari and Ruapehu.

An engraving showing a surveyor at work in the vicinity of Tongariro and Ruapehu, published in New Zealand and its Physical Geography by Ferdinand von Hochstetter, 1867.

Pioneer surveyor Lawrence Cussen completed the first detailed topographical survey of Tongariro in 1891.

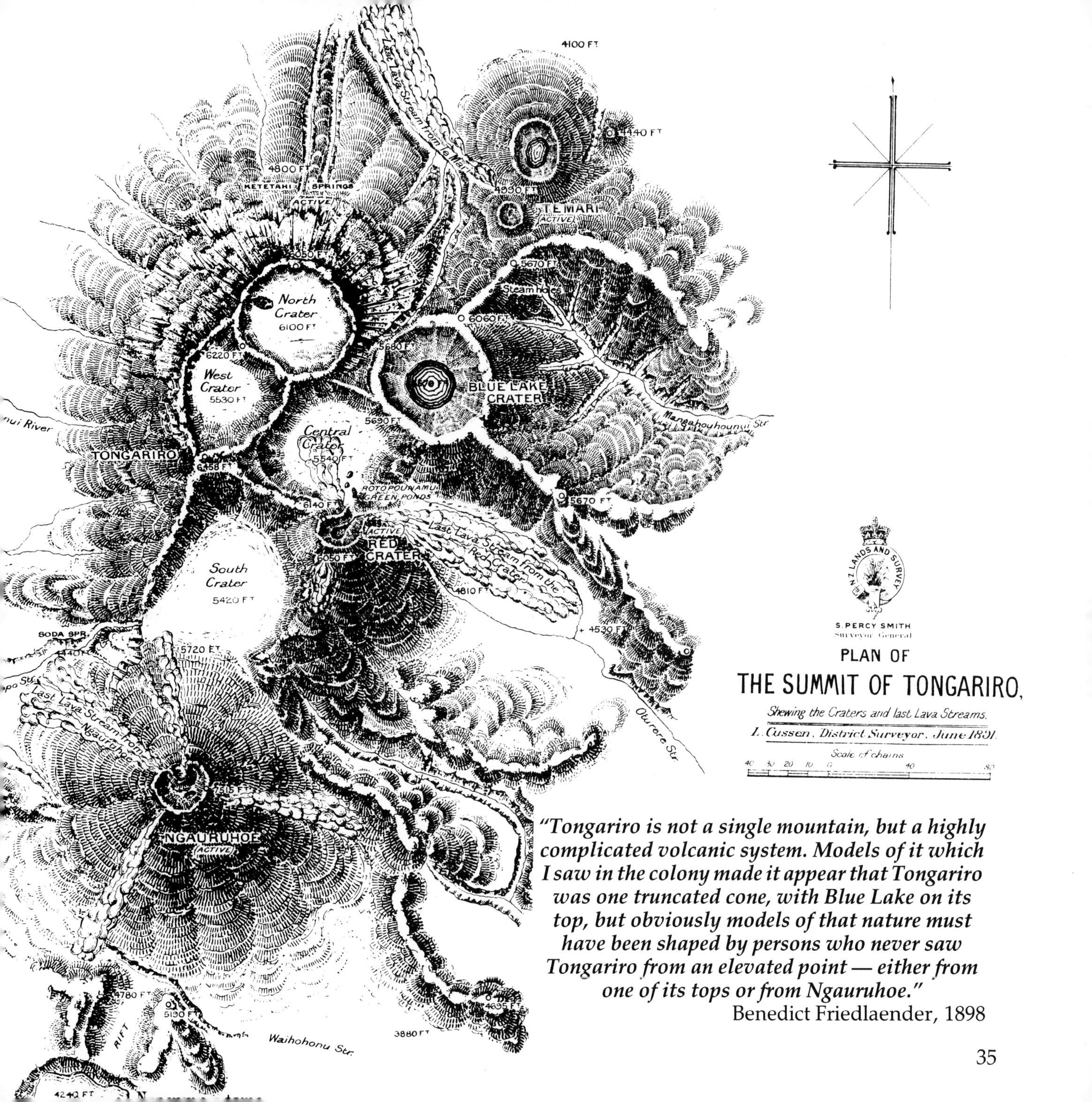

"Tongariro is not a single mountain, but a highly complicated volcanic system. Models of it which I saw in the colony made it appear that Tongariro was one truncated cone, with Blue Lake on its top, but obviously models of that nature must have been shaped by persons who never saw Tongariro from an elevated point — either from one of its tops or from Ngauruhoe."

Benedict Friedlaender, 1898

TE MAARI CRATERS

The Te Maari Craters on the north-eastern slopes of Tongariro have been active for at least 14 000 years and are among the longest-lived of Tongariro's post-glacial vents. Small explosive eruptions occurred here during the multi-vent eruptions from Tongariro around 10 000 years ago. A large lava flow from a vent immediately to the north of the lower crater has an age of between 10 000 and 6000 years.

Eruptions were observed from the lower crater in 1839 and 1867 and from the upper crater in 1868. The eruptions of 1868 were accompanied by violent earthquakes and the area was named after the Maori chieftainess Te Maari who died later the same year. The most recent eruptions took place from the upper crater in 1896/97 when 50mm of volcanic ash fell on the Desert Road and a dusting in Napier.

For more than a century, minor thermal activity has continued in the Te Maari area. Steam vents or 'fumaroles' are found on the walls of both craters.

A group of small explosion pits indent the outer slopes of Lower Te Maari Crater. These unnamed pits are the northernmost site of volcanic activity on Mount Tongariro.

The youngest lava flow in the area stemmed from Upper Te Maari Crater around 500 years ago. A small tongue of andesitic lava flowed into the lower crater but the main flow, numerous overlapping lobes of lava up to 15m thick, cut a 4km-long swathe through dense bush allowing tree-ring dates to be used to pinpoint its age.

NORTH CRATER

The flat-topped cone at the north-western end of the Tongariro massif is a distinctive landmark. A lake of molten lava once filled North Crater and fountaining lava showered the outer slopes of the cone with hot fragments. This lava fused into a layer of rock called "welded airfall" which encases the upper slopes to the north-west and north-east of the cone. Eventually the lava lake cooled, solidifying to form an extensive level surface more than 1000m wide. Only a small remnant of what once must have been an encircling crater rim remains.

The lava lake and the youngest flows are about 14 000 years old. The oldest flows are more than 26 000 years old.

A 300m wide and 40m deep explosion pit in the north-western end of North Crater is surrounded by a ring of debris thrown out in the explosion which created it. Its age is not known although it is older than the Taupo eruption of 1800 years ago.

A fault can be seen cutting across the summit of the cone to the south-east of the explosion pit. Two larger faults traverse the lower western slopes of the cone. Movement has taken place along these faults within the last 1800 years.

CENTRAL CRATER

Despite its name, the depression called Central Crater is not of volcanic origin but occurs at the intersection of the old cone remnants of Tongariro trig and North-East Oturere ridge and the young cones of Red, North and Blue Lake craters.

A thin fan of lava partially covers the floor of the southern half of Central Crater. This black basalt aa flow stemmed from Red Crater less than 1800 years ago. In this photograph, snow emphasises the wrinkled pressure ridges in the surface of the lava which resulted from varying rates of movement in the flow as it moved and solidified.

BLUE LAKE CRATER

Fountains of high-temperature lava were a feature of this vent's euptive history as at nearby North Crater. Red-hot spray from high-temperature lava fountaining to heights of at least 500m has coated a wide area, including the hill of Rotopaunga overlooking Blue Lake. This hard, welded rock layer slowed down erosion on the outer slopes of the crater.

Blue Lake spatter cone is another post-glacial Tongariro vent and was constructed sometime between 25 000 and 10 000 years ago. The vent is now filled by a cold freshwater lake of blue-tinted waters 400m wide and 16.5 m deep. Like the Emerald Lakes its waters are noticeably acidic with a pH of 5.

An aerial view of the Tongariro complex from Red Crater across Central Crater to North and Blue Lake craters.

Blue Lake and North Crater (above right).

Trampers wind their way across the floor of Central Crater toward the northern slopes of Tongariro. Red Crater, Ngauruhoe and Ruapehu form a line to the south.

RED CRATER

Red Crater lies 3km to the north-east of Mount Ngauruhoe and its crater rim forms the highest point, 1886m, on the Tongariro Crossing walk. This spectacular multi-hued crater, 60-80m deep, lies within a scoria cone which rests on top of older Tongariro lava flows. Dark red and black lava, of variably welded basaltic andesite scoria, drapes over the old light grey columnar-jointed lava flows.

A number of small to large lava flows in many directions accompanied the formation of the scoria cone which has been constructed since the last glaciation.

Red Crater has been active quite recently. A series of explosive eruptions have taken place within the last 3000 years from new vents near the centre of the crater. As well, five flows have stemmed from here since the 1800-year-old Taupo eruption. Three flowed into the Oturere Valley and a small basaltic-andesite flow poured down the steep wall of South Crater but did not spread out over the floor of the basin (see photo on page 44). The fifth lava flow was more fluid than the others and left a broad (400m wide) and unusually thin (1-5m) trace. This extends from Red Crater across the floor of Central Crater towards Blue Lake Crater.

The most recent confirmed volcanic activity from Red Crater were several ash eruptions observed between 1855 through into the 1890s, a little over 100 years ago.

Several basaltic-andesite feeder dikes have been exposed by erosion including this remarkable dike on the south-east wall of Red Crater. It was preserved as a hollow lava tube when magma drained from below and fed a lava flow into the south head of Oturere Valley.

It is a steep descent from Red Crater Rim to the beautiful Emerald Lakes. The name Red Crater is derived from the blood-red to chocolate brown basaltic scoria composing its walls. Scoria usually ranges from grey to black in colour when erupted but it is sometimes altered to red by high temperature oxidation of iron in the rock(right).

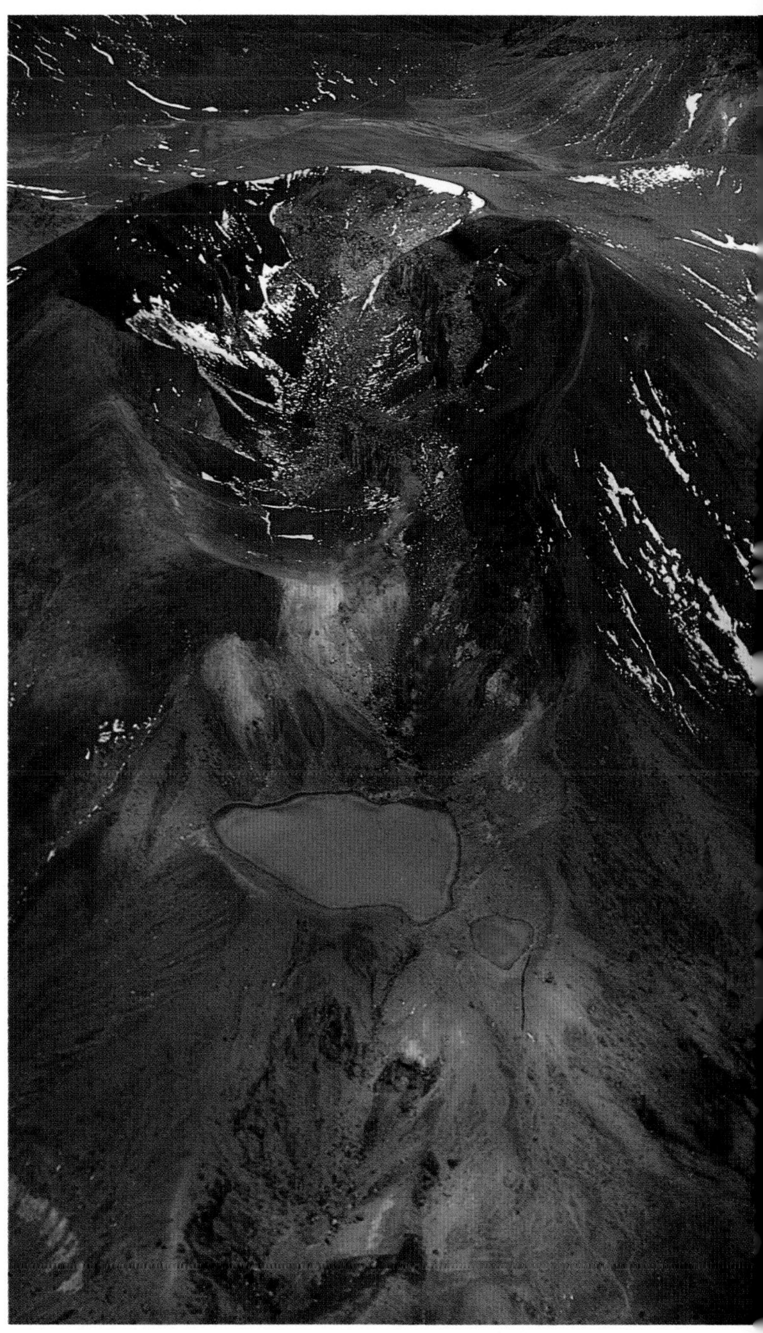

THE EMERALD LAKES

The Emerald Lakes, at the foot of the breached northern wall of Red Crater, fill three explosion pits formed in the past 1800 years. The lakes are up to 4.5m deep and have a pH of 3-5. Although they lie near to warm ground and steaming fumaroles these lakes are cold. Their colour is caused by minerals, mainly fumarolic sulphur, entering the water and forming polysulphide ions.

Surveyor Lawrence Cussen labelled these bright green lakelets "Rotopounamu — the green ponds" on his 1891 map of the area. Today, they are called the Emerald Lakes.

THE OTURERE CENTRE

Evidence of more than 100 000 years of volcanic activity in the Oturere area is preserved in the valley walls. There are at least 70 layers including thick blocky lava flows, deposits of molten airfall material and cemented vent breccia. These layers reveal periods of intense volcanism from several vents, old and young, which have contributed to the growth of the central portion of Tongariro.

North-East Oturere, one of the original Tongariro cones, has been largely removed by glacial erosion. Other younger vents can still be recognised including Half Cone and another small vent at the head of the Oturere Valley.

Post-glacial scoria fall deposits from both North-East Oturere and South-West Oturere have been dated indicating a resurgence of activity from this area around 10 000 years ago.

The pronounced flow pattern of transverse ridges set in the Oturere lava flow is best seen from the air. However, a walk among these weird lumps of jagged lava is a unique experience.

The broad floor of the formerly glaciated Oturere Valley was covered by a large flow of lava between 2500 and 9700 years ago. Erupted from Red Crater, the flow extends for 8km down the valley and is up to at least 30m thick and 400m wide.

An aerial view of South Crater from above the summit of Ngauruhoe. A pool of melt water glistens in the explosion pit within South Crater. The northern peaks of Kakaramea, Tihia and Pihanga can be seen in the distance beyond the Tongariro massif.

SOUTH CRATER

Like Central Crater this large basin is not believed to be a crater at all: instead, its shape is determined by the geography of the surrounding cones. A semi-circular depression at the northern head is probably a glacial cirque.

To complicate matters, long after the glaciation an eruption punched a 100m-wide explosion crater in the south-east corner of this basin. Tephra layers place it between 2000 and 1800 years old. Some time later, lava from Red Crater flowed down the valley wall to the south of Tongariro summit.

South Crater from the summit of Mount Tongariro (1968m). The young cone of Ngauruhoe rises out of the southern side of this elongated depression. Fresh snow highlights the margins of a large explosion pit and lobes of lava that flowed from Ngauruhoe onto the floor of South Crater in prehistoric time.

MOUNT NGAURUHOE

"Ngauruhoe, one of the prominent features of the volcanic chain, is a beautifully symmetrical cone, rising in almost sugar-loaf shape from amidst the ruins of former cones."

Lawrence Cussen, 1891.

About 2500 years ago, a new vent, the youngest within the 275 000 year old Tongariro composite volcano, began to form. Over the centuries, intermittent but fairly frequent eruptions of pyroclastic material and molten lava added breadth, height and bulk to the fast-growing cone of Ngauruhoe. Today, an imposing 2287m peak rises 900m above the surrounding landscape.

The mountain is composed of interleaved layers of ash and lava ranging in composition from andesite to basaltic andesite. Eruptions from a central crater have constructed the steep 30-degree outer slopes of the cone. Constructional processes continue to dominate over erosional processes with regular eruptions (with repose periods of years to decades) maintaining the mountain's conical form.

Ngauruhoe was mildly active for two decades after the eruptions of the mid-1950s and a plume from powerful fumaroles in the crater was usually seen above the summit. Since the eruptions of the mid-1970s Ngauruhoe has experienced the longest recorded period of quiescence since observations began in 1839.

An engraving of Mount Ngauruhoe published in "The King Country" by J.H. Kerry-Nicholls, 1884.

Tall eruption clouds laden with fine volcanic ash formed towering eruption columns above Mount Ngauruhoe in 1974-75. These enormous clouds reached about 13km above the summit.

ASH ERUPTIONS FROM NGAURUHOE

Pyroclastic eruptions, often referred to as ash eruptions, have been the most common kind of eruption from Ngauruhoe since regular observations began about 150 years ago. More than 70 ash eruptions have occurred during this period. Many of these eruptions were small but several phases of violent activity have taken place.

Ash eruptions frequently result when rising magma within the volcano encounters a plug of solidified lava blocking the vent. A build-up of gas pressure beneath the plug eventually leads to an explosion, or series of explosions, as the obstruction is torn apart by the expanding gases. Quantities of ash and blocks of lava are ejected in these eruptions.

Highly explosive ash eruptions took place most recently from Ngauruhoe in January and March 1974, and in February 1975. These eruptions were of a type known as "vulcanian". Cauliflower-shaped eruption clouds and ash avalanches are characteristic features of this kind of eruption.

A major eruption occurred from Ngauruhoe on 19 February 1975. Gas streamed from the crater for several hours producing an eruption plume that was between 11km and 13km high. Numerous pyroclastic avalanches were generated as the cloud collapsed. The avalanches consisted of a turbulent mixture of pyroclastics (ash, bombs and larger blocks) which rolled swiftly down Ngauruhoe, leaving sheets of debris at the foot of the cone.

These ash avalanches have been likened to the devastating "nuées ardentes", or glowing clouds, generated from volcanoes in other parts of the world. In 1902, a nuée ardente from Mont Pelée swept through the Caribbean town of St Pierre 7km away, killing 30 000 people. This glowing cloud was estimated to have travelled at a velocity of at least 160km per hour, and to have had a temperature of 700-1000°C. In contrast, the ash avalanches from Ngauruhoe in 1974-75 were of much lower temperature (about 100°C), moved more slowly (about 60km per hour), and did not extend beyond the base of the cone (2km from the summit).

Molten lava rose to within 50m of the rim of Ngauruhoe's active vent in 1975. At times, fountaining incandescent scoria was visible at the base of the eruption column. No lava flowed from the crater but avalanches of pyroclastic debris swept down the slopes of the cone on several occasions.

A series of seven, highly explosive, individual eruptions followed the continuous gas-streaming eruption of Ngauruhoe on 19 February 1975. The explosions took place at 20-60 minute intervals for more than five hours.

The eruptions were accompanied by atmospheric shockwaves, "flashing arcs", which passed through the air above the vent. Each shockwave was closely followed by the eruption of a compact cloud of highly compressed gas and solid ejecta. Atmospheric shockwaves were also observed during explosive activity of Ngauruhoe in 1954 and 1974.

Cannon-like explosions were heard more than 80km from Ngauruhoe. The eruptions blew chunks of lava (up to 20m across) high above the crater. Lava blocks from these eruptions were scattered within a radius of 3km from the summit of Ngauruhoe.

Explosive eruptions in 1974 and 1975 effortlessly tossed car-sized blocks of lava several hundred metres above the crater of Ngauruhoe. Airborne blocks and bombs were visible from Turangi, 24km away.

This huge block of lava weighing 3000 tonnes was thrown 100m during an eruption of Ngauruhoe in March 1974. Ejected blocks often shattered when they hit the ground and formed "impact craters".

LAVA ERUPTIONS FROM NGAURUHOE

Lava has flowed from Ngauruhoe during three separate eruptive phases since recorded observations began. In July 1870, two flows spilled down the northern flanks of the volcano. Again, in February 1949, lava flowed over the lowest part of the crater rim and moved down the north-west side of the cone towards the Mangatepopo Valley. At least ten separate andesite flows were erupted from Ngauruhoe between June and September 1954. An estimated eight million cubic metres of lava was erupted.

At least 10 separate andesite flows were erupted from Ngauruhoe between June and September 1954. Lava flows piled up in the saddle between Ngauruhoe and Pukekaikiore and tongues of lava flowed onto the floor of the Mangatepopo Valley.

Lava flowing from the crater of Ngauruhoe on 16 September 1954.

Scientists watch the slowly advancing front of a lava flow on 18 August 1954. The lava was relatively viscous and moved at a rate of about 20cm per minute. Some of the flows were more than 18m thick and were still warm almost a year after being erupted.

The andesite flows from Ngauruhoe are all similar — while molten they lost volcanic gases in explosive bursts and cooled to form a jagged, angular surface. This type of lava is known by the Hawaiian term "aa" (pronounced ah-ah), which refers to rough lava on which it is difficult to walk barefoot.

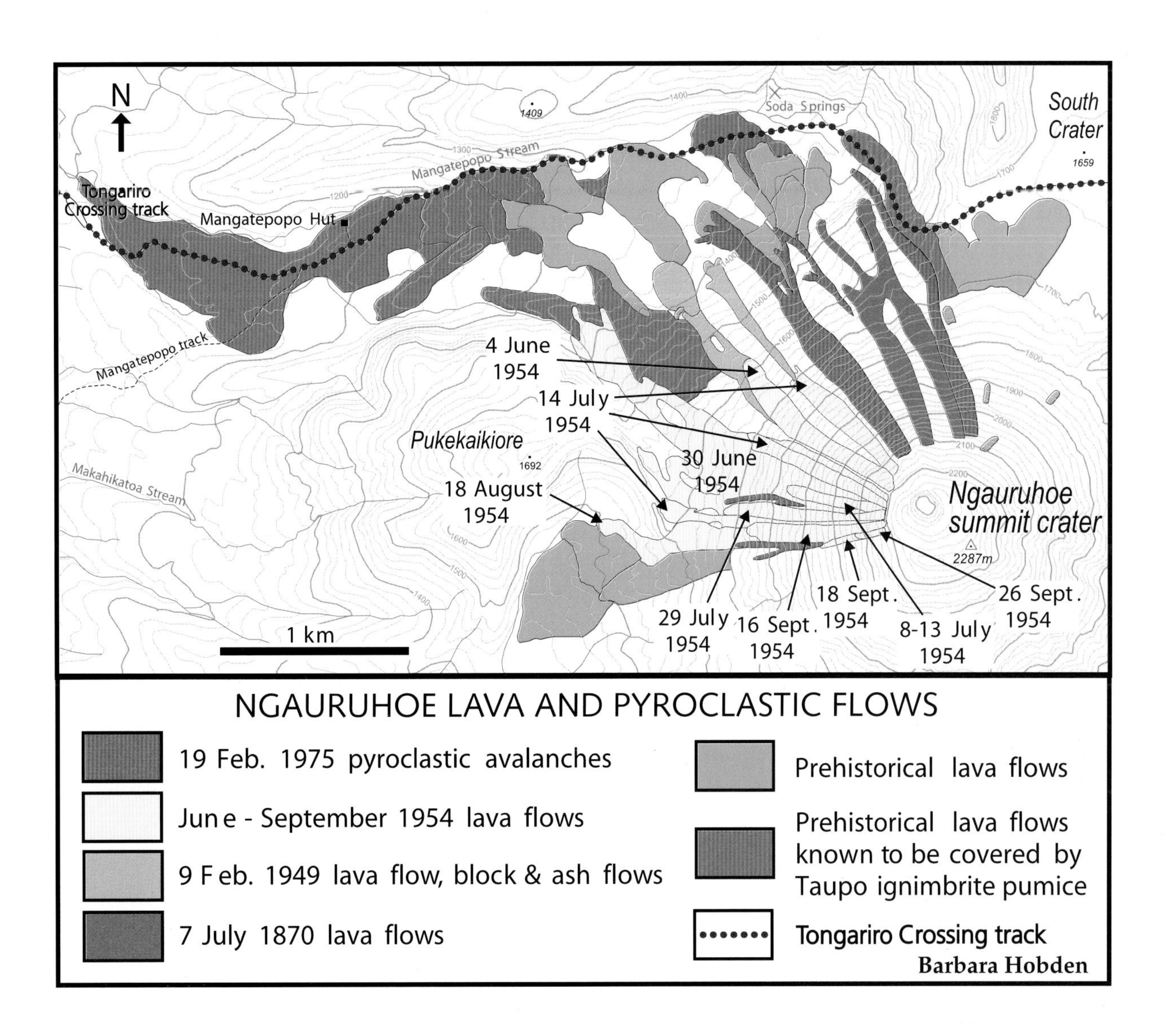

Prehistoric lava flows, some overlain by the 1800-year-old Taupo ignimbrite, have been mapped at Ngauruhoe. Their relief is generally more subdued and they are more vegetated than the youthful flows mapped on the north to north-west side of Ngauruhoe. These modern flows comprise two lava flows from 1870, one lava flow and two pyroclastic avalanches from 1949, at least ten lava flows from 1954 and about five pyroclastic avalanches from 1975.

Pre-existing topography affects flow patterns as demonstrated by the 1870 flows which joined and then diverged around a prominent old Tongariro lava bluff causing the steep flow fronts to break up into loose blocks and scree. Lava flow paths were also diverted by the cliffs of Pukekaikiore. Towards the base of the cone, the young aa lava flows often thickened to become more blocky as a result of decreased gas content and temperature, and increased crystallinity.

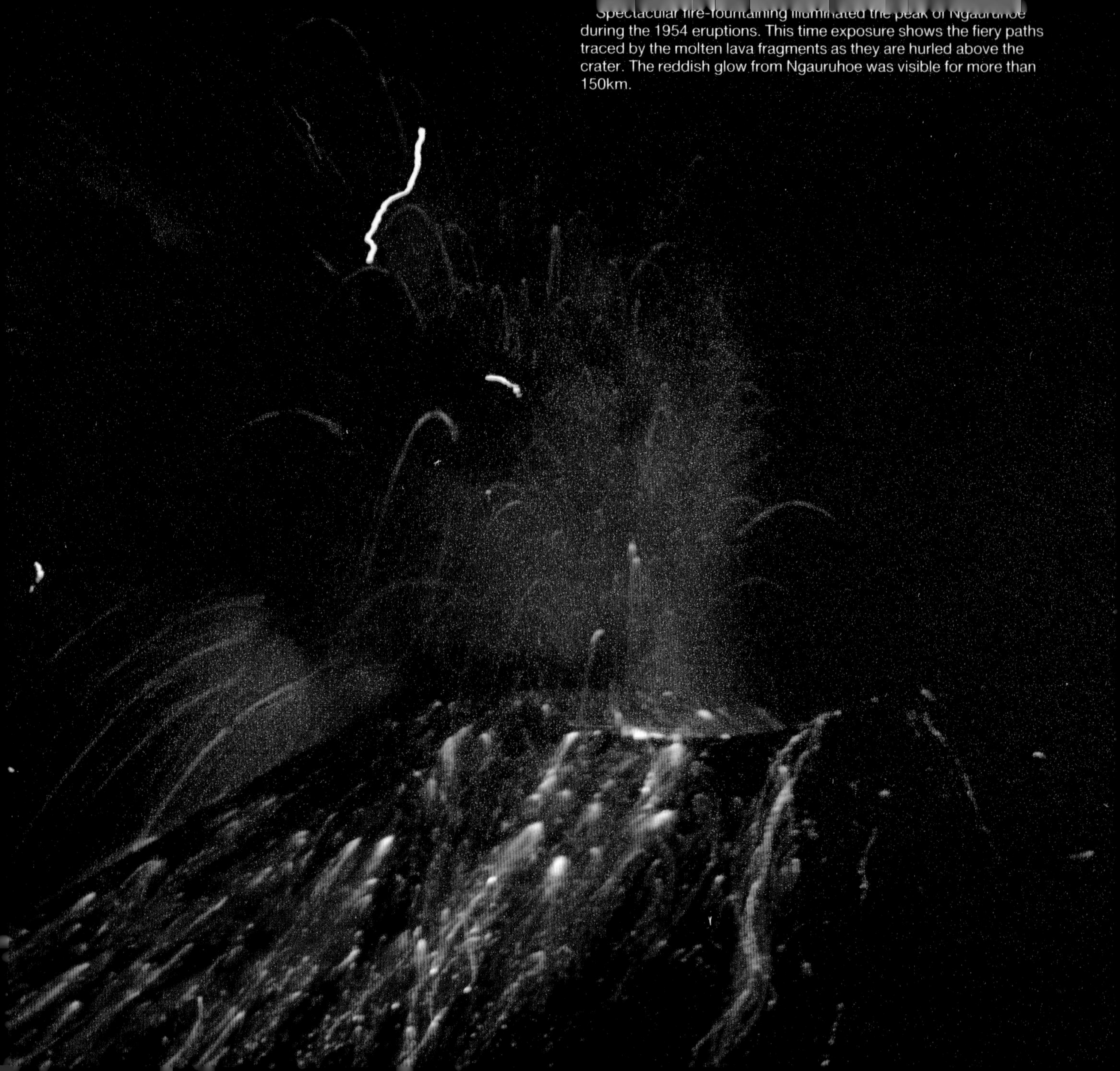

Spectacular fire-fountaining illuminated the peak of Ngauruhoe during the 1954 eruptions. This time exposure shows the fiery paths traced by the molten lava fragments as they are hurled above the crater. The reddish glow from Ngauruhoe was visible for more than 150km.

STROMBOLIAN ERUPTIONS

The emission of lava from Ngauruhoe in 1954 was accompanied by spectacular lava-fountaining. Eruptions blew glowing fragments of molten lava to heights of several hundred metres above the volcano. A much hotter and more fluid andesite magma was involved in these eruptions than in the vulcanian eruptions of 1974-75. The temperature of the lava in the vent is thought to have been about 1000-1100°C.

This type of activity is called "strombolian" after the volcanic island of Stromboli which is known as the "lighthouse of the Mediterranean". Strombolian activity is milder than the explosive vulcanian-style eruptions and isn't considered to be particularly dangerous, except to people within the immediate vicinity of the volcano.

Most of the incandescent fragments ejected in the 1954 eruptions landed on the summit area of Ngauruhoe. Blobs of molten lava rained back down to build a cone around the active vent. Called a "spatter cone", it grew to a height of 60m and completely changed the outline of Ngauruhoe's summit.

NGAURUHOE IN TRANSITION

"The crater was the most terrific abyss I ever looked into or imagined. The rocks overhung it on all sides and it was not possible to see above ten yards into it from the quantity of steam which it was continually discharging. It was impossible to get on the inside of the crater as all sides I saw were, if not quite precipitous, actually overhanging, so as to make it very disagreeable to look over them."

John Bidwill, 1839

A watercolour of the crater of Ngauruhoe in 1876 sketched on the spot by Kennett Watkins.

The major alterations to the summit of Ngauruhoe since Bidwill's visit are illustrated in a series of drawings from the north-west.

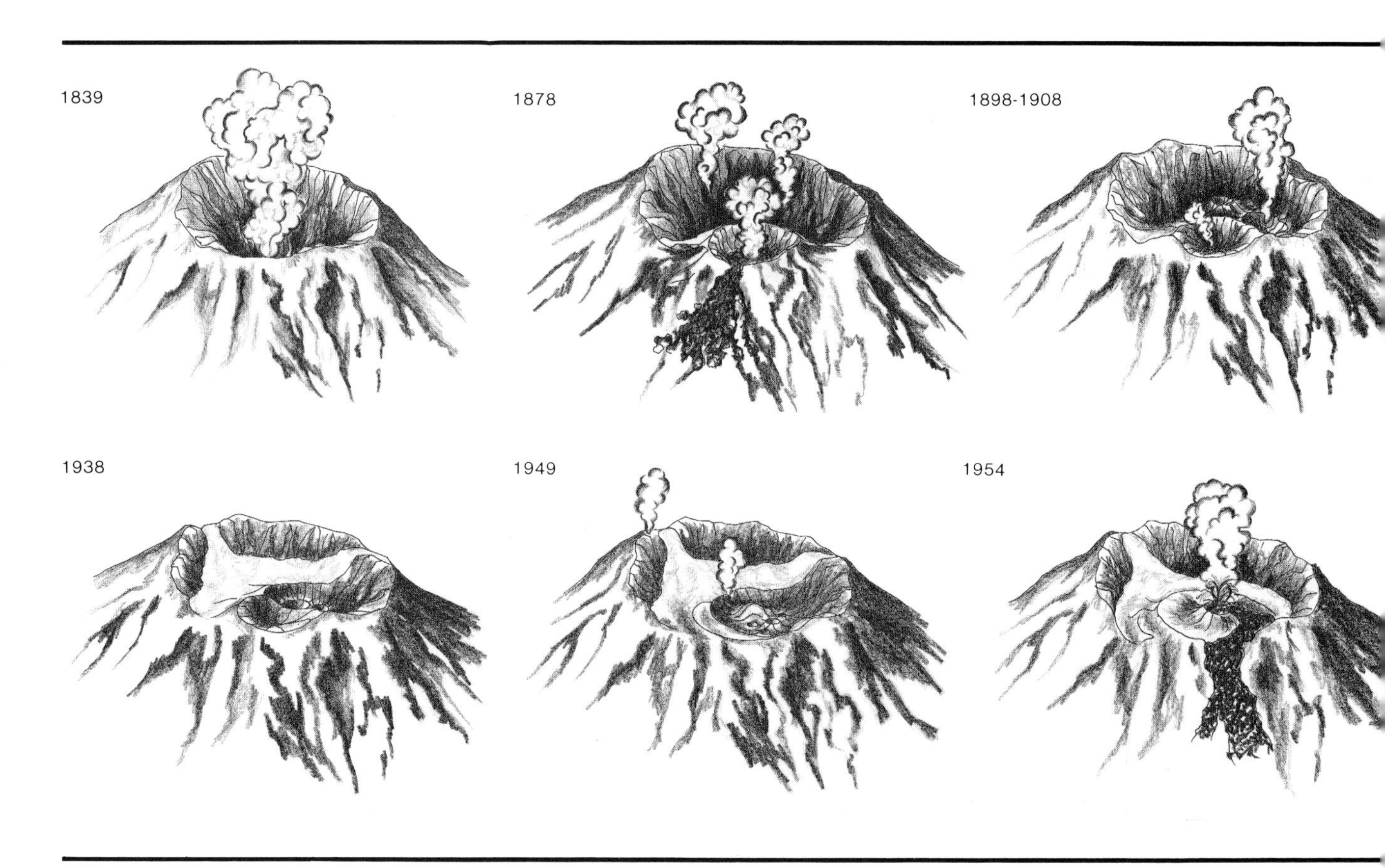

In 1839, when explorer-botanist John Bidwill became the first European to ascend Ngauruhoe, he described a single, large crater with sheer walls. Since then, the summit area has changed dramatically owing to eruptions, infilling and collapse. Different vents have formed and disappeared. As this young volcano continues to evolve, it is likely there will be more major changes to its summit configuration.

Ngauruhoe in July 1984 showing the 200m-wide and 70m-deep active vent nested within the 400m-wide older summit crater. During the eruptive period of late 1973 to early 1975 a plug of lava rose in the vent and pyroclastic ejecta partially filled the active crater. Part of the north-west crater wall also collapsed into the vent which had been more than 180m deep with almost vertical sides prior to these eruptions. Minor collapses and rock falls have occurred since 1975.

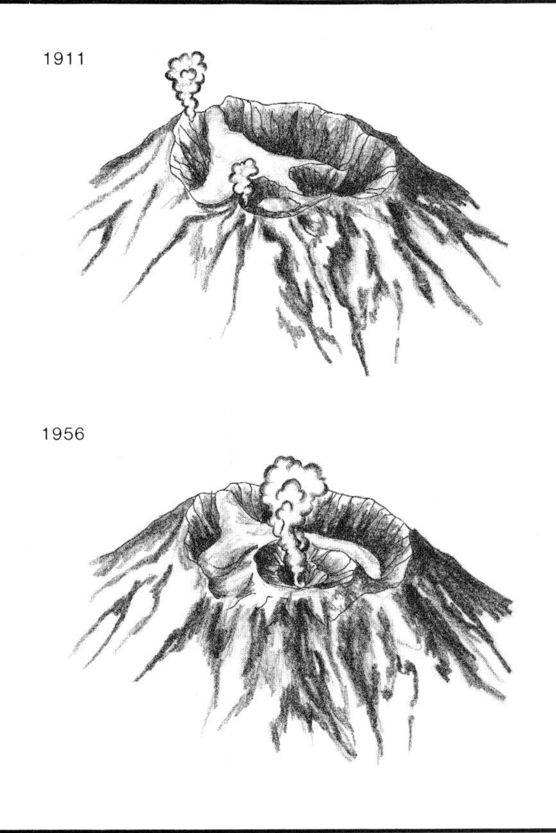

EVEN MOUNTAINS ARE TRANSIENT FORMS

Ngauruhoe is one of the most striking landforms in Tongariro National Park, yet as a feature of the landscape this peak has only been around for a short time geologically - 2500 years, according to radiocarbon dates of vegetation killed by the earliest showers of Ngauruhoe ash.

It was previously thought that a similar, but larger, single ancestral peak, referred to as 'proto-Ngauruhoe', existed in its place. However, Potassium-Argon dating has recently shown that the remnant ridges radiating out from the volcano are in fact of different ages. This suggests 'proto-Ngauruhoe' may in fact have looked more like Ruapehu, a mountain that has been built from eruptions from a number of different vents.

Large Ice Age glaciers carved the Mangatepopo Valley, Oturere, Waihohonu and other valleys in the area and have been credited with the destruction of this earlier proto-volcano.

Mount Ngauruhoe.

Evolution of a mountain — a series of sketches illustrating the type of changes which may have preceded Ngauruhoe as we know it today.

260 000	A peak broader and probably higher than present day Ngauruhoe once existed on the southern part of Tongariro. 260 000-year-old lava flows near Tama Lakes give an approximate age for this peak.
100 000	With the onset of the Ice Age, a cap of permanent snow and ice formed on the summit of Proto-Ngauruhoe, and glaciers began to develop.
50 000	At their maximum, the glaciers were up to 5km long. Glacial erosion cut into valley heads and deepened existing gullies, slowly wearing down the height and mass of the old volcano.
14 000	As the world's climate warmed, the glaciers began to retreat and eventually to disappear. "Shoulders" of resistant lava were left separated by deeply glaciated valleys.
2 500	Volcanic activity resumed and a new cone began to grow on top of the dissected old volcano. Eruptions of lava and tephra started to fill the glacial valleys, gradually burying the glaciated relief.

At left - Today, an unglaciated volcano rises above the eroded remains of proto-Ngauruhoe. Ten thousand years ago there were several very active vents in this area. Some of these vents are now buried while others are still exposed including Tama Lakes (see centre bottom) and Half Cone to the north-east of Ngauruhoe in the upper Oturere Valley - look for the large horseshoe-shaped feature in top right sector of photo south of Red Crater. Lava flows on Ngauruhoe and in Oturere Valley are also very prominent.

PUKEKAIKIORE

The eruptions which built Ngauruhoe during the past 2500 years have buried several early cones of the Tongariro volcanic complex. However, the eroded remnants of another old cone survive immediately to the west of Mount Ngauruhoe on the south side of the Mangatepopo Valley. Pukekaikiore (1737m) was active between 190 000 and 120 000 years ago. A source, now concealed beneath Ngauruhoe, produced thick columnar-jointed silicic andesite flows.

Pukekaikiore was eroded and modified by Ice Age glaciers which occupied the Mangatepopo and Makahikatoa valleys during the Holocene period. Activity resumed around 15 000 years ago. This produced olivine-andesite lava flows on the western side and a scoria cone near the summit.

PUKEONAKE CONE AND ASSOCIATED VENTS

The tussock and shrub-covered scoria cone of Pukeonake is located 5.6km to the west of Ngauruhoe. This 140m high basaltic-andesite cone was formed by eruptions from twin vents. Its crater is filled with Oruanui tephra and is, therefore, more than 23 000 years old.

Another two smaller residual cones to the north of Pukeonake are also part of this satellite centre of Tongariro which developed along a north-south fissure parallel to faults in the area. Extensive lava flows from this fissure fill valleys to the west and cover an area of 55 square kilometres. The oldest flows are exposed at the Mahuia Rapids. Younger flows outcrop along the Pukeonake, Papamanuka and Whakapapanui streams. They represent the only known eruptions peripheral to the main Tongariro complex.

Since Pukeonake's formation, stream erosion and the draining of a crater lake have cut a steep-sided gully inside the cone, eroding away the south-west crater wall. Erosion is even more advanced on the other two small vents which are less than 20m high. Perhaps they lay in the path of a glacial outwash stream from the Mangatepopo Valley that stripped them of their tephra cover.

TAMA LAKES AND CRATERS

Ancient lava flows and post-glacial eruption products reveal the Tama Lakes area has been an important centre in the history of the Tongariro complex. Activity has occurred from several andesite cones, including vents identified as Tama 1 and 2. Remnants of the 200 000-year-old Tama 2 cone can be seen in prominent ridges forming the glaciated walls of Waihohonu Valley. The earliest lava flows (275 000 years old) found on Tongariro are exposed in outcrops around the two northern craters adjoining Lower Tama Lake.

In contrast, the craters containing the Tama Lakes are much younger post-glacial features. Surge deposits and breccias mantle the area around these explosion craters and dacite pumice can be found nearby. Further activity occurred to the south of the Tama Lakes less than 10 000 years ago when explosive eruptions blasted welded scoria from new vents.

Upper Tama Lake fills a depression formed by explosive eruptions from at least two vents.

Nearby, Lower Tama Lake occupies the lowest of three adjoining explosion craters. Volcanic debris is being washed into the complex from above and the sediments are beginning to affect the lake.

(Far right), Nga Puna a Tama - "The Springs of Tama" were named after Tamatea, the high chief of the Takitimu Canoe, who explored this area more than six centuries ago.

SADDLE CONE

Evidence gathered from pyroclastic deposits in the Tama Saddle area reveals several vents, now buried, were active in this area about 10 000 years ago. However, two small centres can still be recognised about 3km south-west of Lower Tama Lake. Arrows show the location of Saddle Cone at 1550m and a cross marks the other vent site at 1900m on the northern slopes of Ruapehu. Lava flows from Saddle Cone's small crater are probably also around 10 000 years old. A small block lava flow from the vent at 1990m may be significantly younger.

Arrows pinpoint the location of "Saddle Cone". Its crater floor is 50m wide and lava has breached the north-west wall of the cone (right).

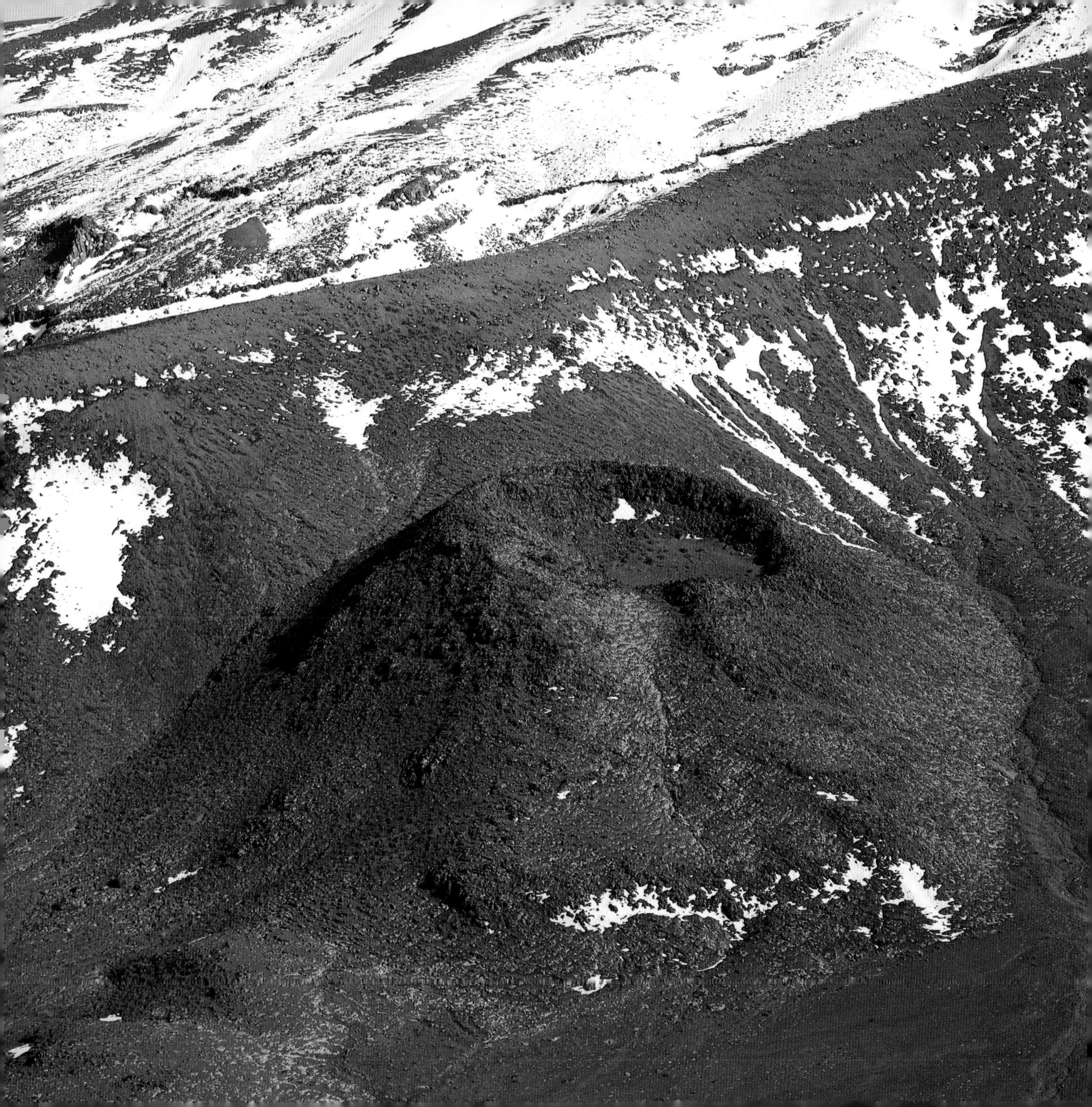

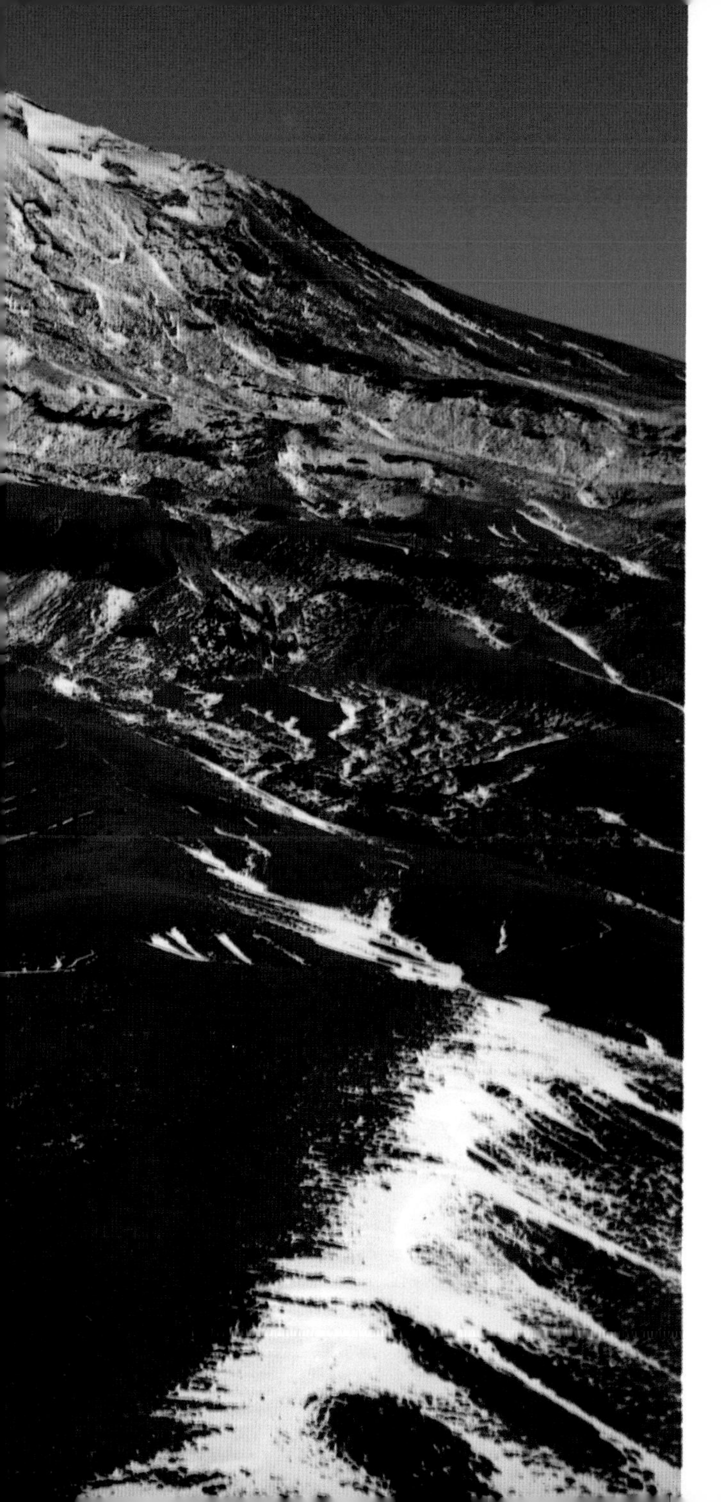

THE RUAPEHU MASSIF

The imposing andesite-dacite stratovolcano of Ruapehu lies near the southern end of the Taupo Volcanic Zone. Its 2797m summit is the highest point in the North Island and the mountain itself is even larger than Tongariro.

At least four distinct cone-building episodes involving both summit and flank vents have been identified during the past 250 000 years, each lasting perhaps tens to hundreds of thousands of years. This is said to be typical of composite arc volcanoes which tend to grow in spurts.

Tephra studies indicate the present active vent has been the focus of activity for least 2500 years. In the past 120 years, dozens of steam eruptions have been reported from the active crater which is usually filled by a lake. Small-volume magmatic episodes occur less frequently having a periodicity of 20-30 years in historic time.

This magnificent volcanic landform has also been modified by the processes of erosion and, in particular, the last glaciation wrought major changes at Ruapehu.

"Mount Ruapehu, the apex of the North Island... grandly crowns the Tongariro National Park. Its contour, especially viewed from the North, certainly resembles a great crown, sweeping up from the domed centre of the island to a supernal series of glittering points of ice-fire."

James Cowan, 1927

TE HERENGA CONE

The long jagged rib of Pinnacle Ridge, on the northwest slopes of Ruapehu, is an eroded remnant of an old andesite cone given the name Te Herenga. Lava flows from this ancient cone have been dated at 230 000 years and are the oldest mapped rocks at Ruapehu.

Glacial and stream erosion destroyed Te Herenga Cone but its heart, an underlying network of dikes, magma intrusions, hydrothermally altered lava flows and breccias of the Te Herenga Formation, has been laid bare along the upper part of Pinnacle Ridge.

Similarly, Mead's Wall, near the base of the Whakapapa Skifield, is a vertical volcanic dike that has been exposed by erosion.

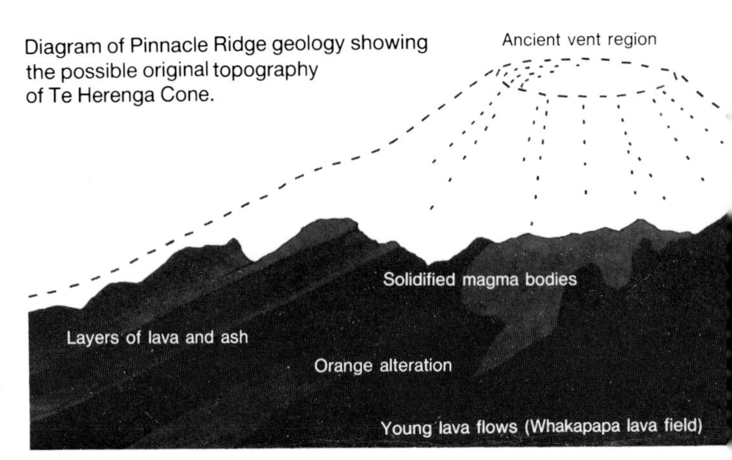

Diagram of Pinnacle Ridge geology showing the possible original topography of Te Herenga Cone.

Pinnacle Ridge, Mount Ruapehu.

WHAKAPAPA LAVA FLOWS

Dates obtained from tephra layers indicate that the Whakapapa lava flows were deposited between 5000 and 9700 years ago. The flows may have originated from "Dome Crater", a vent on the Summit Plateau of Ruapehu.

An estimated volume of 1.5 cubic kilometres of lava poured down the Whakapapanui Valley for about 8km. Massive flows of blocky lava partly filled the glacial valley that had been cut long ago through the ancient cone of Te Herenga. The Iwikau Ski Village and the Whakapapa Skifield are situated on these young flows.

Chaotic masses of large, dense, angular blocks mixed with smaller material are exposed in cuttings of the Bruce Road en route to the Whakapapa Skifield. These red oxidised deposits, called "autoclastic breccias", formed on the outer margins of the Whakapapa flows and were produced by the crumbling of flowing lava.

It is thought that some of this material was immobilised on the steep upper slopes of the mountain and that avalanches of hot tephra were generated. These flowed along pre-existing stream valleys and extended almost 20km. The upper portion of the lahar mounds near the Chateau may be avalanche material associated with the Whakapapa lava flows.

CRATERS OF THE SUMMIT REGION

The summit region of Ruapehu stretches 3km from north to south and is up to 1.5km wide. Eruptions from multiple vents and glacial erosion are largely responsible for the shape of the broad and irregular summit of today. Snow and ice 'permanently' mantle much of the mountain-top making it difficult to interpret the subglacial topography. However, a radio-echo survey indicated ice depths of at least 130m at the north-eastern end of the Summit Plateau.

It was once thought that a single large crater had been active between Te Heuheu Peak and The Dome. However, a substantial thinning of the snow and ice cover since the beginning of last century has exposed two overlapping craters in this area - North Crater and Dome Crater.

Present-day activity has been modulated through a vent system located beneath Crater Lake at the south-west end of Ruapehu's extensive summit region.

The outline of the ice-filled North and Dome craters can be distinguished on the Summit Plateau of Mount Ruapehu. The remains of another crater, East Crater, are believed to form the upper part of the Whangaehu Glacier. A fourth vent, known as both Girdlestone Crater and Wahianoa Cone, was active for a time between Mitre and Girdlestone. Mitre Peak is an erosion remnant of this cone.

Crater Lake occupies the present active crater which lies in a volcanic cone near the eastern edge of West Crater.

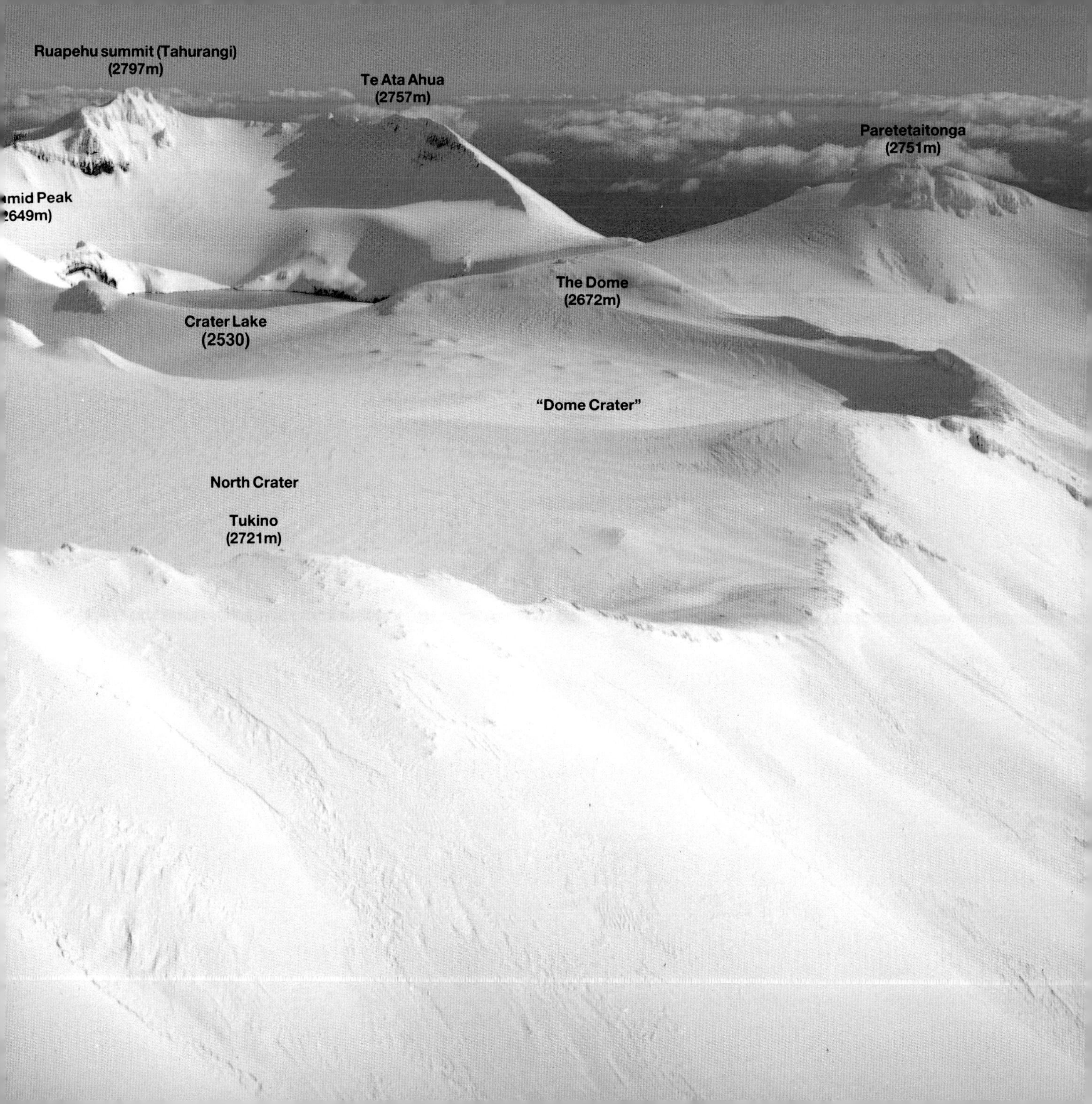

Ruapehu summit (Tahurangi)
(2797m)
Te Ata Ahua
(2757m)
Paretetaitonga
(2751m)
amid Peak
2649m)
The Dome
(2672m)
Crater Lake
(2530)
"Dome Crater"
North Crater
Tukino
(2721m)

An old photo of two climbers beside Crater Lake probably taken around the turn of the century.

The first recorded eruption of Mount Ruapehu was reported in the "New Zealand Spectator and Cook's Strait Guardian" in 1861: *"Ruapaho Volcano — This volcano which has been quiescent for a period beyond the recollection of the oldest inhabitant, suddenly burst forth into full activity on the 16th ult., [16 May 1861] when dense volumes of smoke were seen issuing from the crater, and at night a lurid glare was reflected by the heavens, considerably astonishing the natives."* Despite this eruption, it was still widely believed that Ruapehu was extinct.

The first climb on a summit spur of Ruapehu was made in 1853, but it was not until 1879 that the crater lake was discovered. George Beetham and Joseph Maxwell were the first to see the lake, however there was no indication that the water was warm on that occasion. James Park reported in 1886 that the site of Crater Lake *"was occupied by a great sheet of ice, of a bluish colour, and there was no appearance of steam or water."*

Only a few months later, Government Surveyor Lawrence Cussen found the lake in a different condition: *"I noticed little clouds of steam rising from the surface of the water. On watching more closely, the water appeared now and again to assume a rotatory movement, eddies and whirlpools passing through it from the centre to the sides, and steam flashing up from those eddies, leaving little doubt in my mind that the water was in a boiling state."*

The existence of the hot summit lake on Ruapehu was of great interest because it provided direct evidence of volcanic activity. A violent eruption in 1895 from Crater Lake, which produced a cloud of steam seen from a considerable part of the North Island, confirmed that Ruapehu was indeed an active volcano.

"On Sunday morning March 10, 1895, the snow-capped giant Ruapehu was seen . . . to rouse from his long slumber, and to lift the snowy nightcap which for so many years rested peacefully on his head. Presently it was evident that the giant was fairly awake, and from his lofty summit there rose in slow and majestic grandeur a magnificent cloud."

Josiah Martin, 1895

Ruapehu in spectacular eruption on 10 March 1895 (drawing by unknown artist). The eruption cloud is an estimated 2600m high (8-9000 feet). Mount Ngauruhoe, the Te Maari Craters and Ketetahi Springs are also steaming vigorously.

THE ACTIVE CRATER OF RUAPEHU

A crater lake, with a diameter of about 500m, normally fills the active crater of Mount Ruapehu at an elevation of 2530m. It provides a remarkable contrast between volcanic heat and glacial ice. The lake is usually warm, sometimes reaching temperatures of 60°C or more.

Ruapehu's active vent was emptied of water twice last century - during the eruptions of 1945 and 1995/96.

Recent evidence suggests this lake has been a feature of the crater for the last 900 years but volcanic activity has been centred here for about 2000 years.

Panorama of Crater Lake, May 1982.

CRATER LAKE TEMPERATURE

Crater Lake usually has a temperature of between 20° C and 40° C. The highest temperature measured is 60° C and the lowest 10.5° C, although the lake actually froze during the winters of 1886 and 1926.

The water is heated by volcanic gases (given off from magma) as they rise up through the lake. Seismic evidence suggests there are a number of magma bodies 2-9km beneath Ruapehu.

Higher temperatures are associated with an increase in volcanic activity and in the amount of magmatic gas injected into the lake by lake floor fumaroles. Low lake temperatures do not mean that activity has ceased. Rather, they suggest that the vent is blocked. This may be caused by magma beneath the lake congealing and forming a temporary barrier to the hot fumarolic gases. Thus, the water temperature of Crater Lake can provide valuable information about volcanic activity in the vent beneath it.

Diagram showing a cross-section of Crater Lake.

LAKE WATER COMPOSITION

The pristine snow and icefields surrounding Crater Lake are the source of the lake water, yet its composition is highly acidic (pH of 0.8 — 1.5). The lake is also rich in dissolved solids (sulphate, chloride, and lesser amounts of magnesium, calcium, sodium, iron, silica and other compounds).

The chemical constituents of the water are derived from volcanic material, mainly through two processes. Underwater fumaroles containing strong acids supply volcanic gas to the lake water. Minerals are then added to the water through the disintegration of hot andesite rock interacting with the acidic lake water.

Some of the material settles out to produce a bottom layer of very fine, grey mud. During active periods, the turbulent ascent of gas from the magma disturbs the mud layer, turning the lake a battleship-grey colour.

Ascending gases and convecting hot water sometimes carry up sulphur which collects on the surface of the lake as yellow or greenish-black slicks. Some scientists consider that a pool, or pools, of liquid sulphur underlies the mud bottom of the lake in places.

THE OUTLET

Water overflows from Crater Lake via a natural stream and ice cave outlet into the Whangaehu River at the foot of the south-west branch of the Whangaehu Glacier on the eastern side of Ruapehu. The water flows out at a low point in the crater rim on the southern shores of the lake. A steady seepage of lake water also occurs through the crater walls ensuring that the Whangaehu River remains contaminated even when the level of Crater Lake is below the outlet. Consequently, the river is naturally polluted and has no fish and little insect life in its upper reaches.

During quiet periods the waters of Crater Lake range from a pale blue to a deep green. However, even mild volcanic activity increases the amount of suspended material in the water and turns the lake a metallic grey colour. Although it is possible to swim in the acidic waters of Crater Lake, apparently without harmful effect, it is not encouraged. Frequent eruptions from the lake have shown that its mood can change rapidly and with little warning.

ERUPTIONS FROM RUAPEHU

The eruptive activity of Ruapehu in historic time has been largely influenced by the presence of the lake in the active crater. Minor steam eruptions, sometimes called "phreatic" or "hydrothermal" eruptions occur from time to time in Crater Lake. These geyser-like eruptions happen when superheated water flashes to steam, raising the water at the surface of the lake in the form of a dome. The heating process may be caused by conductive transfer of heat to the crater from underlying lava, or by the injection of steam rising from magma. No fresh ash or magmatic debris is erupted.

Major "phreatomagmatic" eruptions are sometimes initiated beneath Crater Lake through the interaction of lake water and molten magma. These eruptions are "surtseyian" in style — taking their name from the submarine eruptions which formed the volcanic island of Surtsey off the coast of Iceland in 1963.

A hydrothermal eruption begins to bubble up in the centre of Crater Lake on 2 January 1982. Moments later, a column of steam was ejected to a height of about 300m above the lake.

Phreatomagmatic eruptions have occurred intermittently from Ruapehu in European times. These explosive eruptions take place when magma rising in the vent encounters wet sediments on the lake bottom or large volumes of lake water. The molten magma chills, contracts and shatters. A chain of violent explosions follow as water and magma continue to interact.

A steam-rich eruption cloud may rapidly rise vertically to hundreds of metres above the crater. Water, lake sediments and fresh lava fragments may surge outwards from the base of the eruption cloud, onto the icefield surrounding Crater Lake. This debris can cause rapid melting of ice and snow, and is often remobilised as volcanic mudflows (lahars).

This sequence of photographs of a phreatomagmatic eruption was taken on 8 May 1971. An explosion originating about 1km beneath the lake ejected water, mud, hot ash, and partially molten rocks to a height of 800m. The rocks are travelling at an estimated 100 and 150m per second.

With each explosion, a black ash-filled cloud is erupted into the air. Lava bombs and blocks driven by the super-heated vapour shoot out of the eruption cloud, trailing dark tails of ash. The fragments produce a fan-shaped effect, called a "cock's tail", as they ascend then descend.

THE 1969 ERUPTION

A major phreatomagmatic eruption occurred in 1969 when lava was extruded beneath the floor of Crater Lake. This led to a series of violent eruptions on the night of 22 June 1969.

Local earthquake and eruption vibrations commenced only 29 minutes before the eruptions began. An initial explosion is believed to have sent powerful horizontal blasts of wet, warm ash around the summit and down the north-west slopes of the mountain. A large eruption followed which resulted in the ejection of about 30 percent of the Crater Lake water. Lake sediments, vent debris and a small proportion of fresh andesite pumice were also ejected.

The 1969 eruption coated the summit region of Ruapehu with a thick layer of ash and mud. Blocks and bombs of andesite were scattered within a zone about 100m wide around Crater Lake. The largest block thrown out in the eruptions was close to 4m in diameter.

The 1969 eruptions generated mudflows which flooded down the Whakapapanui, Whakapapaiti, Mangaturuturu and Whangaehu Valleys. It was the first time since 1895 that lahars had spilled down the north-west and south-west slopes of Ruapehu. Mudflows in 1925 and 1953 were confined to the Whangaehu Valley because a low point in the rim enclosing Crater Lake makes this valley the preferential route for lahars.

Ruapehu from the Bruce Road after the June 1969 eruption. A coating of ash darkens the snow in a broad band down the north-western slopes of the mountain.

The Indonesian word "lahar" is used to describe mudflows on volcanoes. These torrential flows of water-saturated volcanic debris move down-slope under the influence of gravity. They get channelled into valleys and can travel at speeds of more than 50 kilometres per hour.

Eruptions may generate mudflows by depositing water or hot debris onto the slopes of a volcano. However, volcanic activity is not the only cause of lahars — the walls of a crater lake may burst or heavy rain saturate loose volcanic debris, setting lahars in motion.

A large phreatomagmatic eruption on 24 April 1975 (equivalent to about 20 kilotons of high explosive) again triggered mudflows down the north, south and eastern slopes of the mountain. The level of Crater Lake fell by more than 8m as nearly a quarter of its water was ejected. The mudflows damaged skifield facilities, bridges and hydro-electric construction on the Tongariro Power Scheme.

The lahars were the final events of the 1969 eruption. They stemmed from water and debris erupted from Crater Lake and from snow and ice removed from the summit area. On the northern slopes of Ruapehu near the top of Knoll Ridge, two small flows joined, swept into the Whakapapanui Valley, and on down through the Whakapapa Skifield.

The mudflow which ran through the Whakapapa Skifield was up to 2.5m thick and 30m wide. A cafeteria lying in its path was wrecked. Had the eruption and subsequent mudflows occurred during the day, the lives of many skiers would have been threatened.

THE 1945 ERUPTION

Volcanic activity of a kind that had not been witnessed before on Ruapehu took place during 1945. In that year, viscous andesite lava rose quietly in the vent gradually displacing the lake. Water poured out through the Crater Lake's natural outlet, an ice cave on the southern side draining into the Whangaehu Valley. Only after the lake had disappeared did the explosive activity commence.

A chain of events which began with this eruptive phase was to lead 8½ years later, to one of the worst disasters in New Zealand's history.

The Lake Disappears

In March 1945, a plug of semi-solid lava was forced up the vent until it emerged at the surface of Crater Lake. The lava dome, or "tholoid", was like a steaming island at first but slowly mushroomed out over the crater floor. By July, the lava had reached the entrance of the outlet ice cave and the crater was almost empty of water.

Violent Ash Eruptions

Between August and November 1945, frequent explosive eruptions took place from the empty crater. Hot lava blocks blown out by the explosions pitted the upper slopes of Ruapehu within 1.5km of the active vent. Tremendous clouds of ash-laden steam were erupted and the mountain top was blackened with a thick layer of ash. Volcanic ash fell as far away as Wellington.

A series of sketches showing the active crater of Ruapehu during the eruption of 1945.

Dramatic Changes in the Crater Floor

By the end of 1945 when the eruptions had ceased, it was found that the floor of the crater had changed dramatically. The central vent had enlarged considerably and explosion pits had formed in the crater floor. In January 1946, large pools of water were noted in the deeper parts of the crater. Volcanologist Jim Healy who climbed Ruapehu towards the end of January 1946 reported: *"The inner crater consists of a steep walled vent about 1,000 feet [300m] deep from the inner rim, and the bottom is occupied by a boiling lake. The water in the boiling lake is a muddy colour and periodically large quantities of steam issue from the surface."*

This composite photograph shows the active crater during this period. In fact, it was only in the latter half of 1945 and early 1946 that the inner crater of Ruapehu has been seen.

Inner crater of Ruapehu, January 1946. Compare this view with the panorama of Crater Lake (on pages 74-75) taken in 1982.

The Crater Begins to Fill Again

From 1946 onwards the lake slowly increased in size. By August 1953, Crater Lake had stabilised at a level 8m higher than before the eruptions of 1945. The lake was able to fill to a higher level because volcanic debris erupted in 1945 had raised the rim of the inner crater. Ice from the glacier in the crater basin had flowed in behind, partially supporting this new barrier. A small ice cave had reformed at the lowest notch in the newly raised rim. Lake water discharged through this ice cave for at least four months prior to 24 December 1953.

THE TANGIWAI DISASTER

The Barrier Collapses Starting a Lahar

The barrier of volcanic debris formed by the 1945 eruption collapsed suddenly at about 8 p.m. on Christmas Eve 1953. The sequence of events immediately preceding the collapse is not known but most likely involved lake water melting glacial ice. While lake water was overflowing through the outlet ice cave, it was probably also seeping through the barrier and beneath the ice further west. This formed a lower channel for the water and melted the supporting ice behind the ash and scoria barrier which eventually had insufficient strength to withstand a head of several metres of lake water, and it collapsed. Warm, swiftly flowing water from the lake rapidly created a second, much larger ice cave beside the first, allowing an enormous volume of water to pour down the channel beneath the ice into the Whangaehu Valley. The level of Crater Lake dropped about 6m in 2½ hours.

There is no evidence that volcanic activity played any part in the collapse of the debris barrier. However, sudden movement or cracking of undermined ice may have contributed to the collapse.

The Bridge is Swept Away

The flood swept down the Whangaehu River Valley picking up huge quantities of sand, silt and boulders. It reached Tangiwai, 30 km away about two hours after the barrier retaining the Crater Lake had burst. The dense flood wave surged through the area soon after 10 p.m. The lahar's rate of flow at its maximum was about 900 cubic metres per second, equivalent to the Waikato River in flood. The deep, fast moving, debris-laden water carried away parts of the Tangiwai Rail Bridge, including two central piers. Only minutes later, the locomotive and six carriages from the Auckland Express plunged into the river at Tangiwai. One hundred and fifty-one people were killed.

The Aftermath at Tangiwai

The raging torrent spread debris over a wide area. Railway carriages were dumped along the banks of the Whangaehu River downstream of the Tangiwai Rail Bridge. A concrete pier weighing 125 tonnes was carried 64m by the lahar.

Water from the lake poured into this large ice cave (about 50m wide and 30m high) which was created in glacier ice on 24 December 1953. This photograph of the outlet cave was taken four days after the Tangiwai disaster.

Tangiwai, December 1953

HAUHUNGATAHI

Some of the earliest eruptions from the Tongariro Volcanic Centre are believed to have occurred from Hauhungatahi, a deeply eroded volcano located 12km to the west of Ruapehu. A broad plateau reaching an apex of 1500m survives today – a remnant of the lower slopes of the original vent.

Lava flows cap thick layers of ash and block and bomb beds which in turn lie on top of siltstones. Vegetation patterns reflect the difference in the underlying rocks – the tussock-covered volcanic plateau merges with the forested sedimentary block at the break in slope on the sides of Hauhungatahi.

The siltstones contain fossilised oyster, scallop and barnacle shells which seem out of place on a volcano. However, the shells were deposited in this area long before volcanic eruptions began and the shells indicate that the rock had its origins in a marine setting during the Tertiary period about 10 million years ago. Sediments of a similar age also outcrop in the Horopito-Ohakune area.

Since Hauhungatahi ceased activity, it has been showered in tephra from the eruptions of Tongariro and Ruapehu. The remains of trees buried by the pyroclastic flow from Taupo in A.D.186 can be seen exposed on the northern slopes of this weathered cone. Hauhungatahi was at the southern limits of this devastating flow and vegetation on its western side was left untouched.

OHAKUNE CRATERS

A cluster of volcanic vents are located on swampy ground near the edge of a large outcrop of marine sediments, north-west of Ohakune. The lava which erupted from these craters came to the surface along faults beside this uplifted sedimentary block.

Violent explosive activity, probably caused by gas-rich magma encountering abundant ground water, blew out blocks and bombs. These were deposited in broad flat rings around a number of small craters, 100-600m in diameter. Coarse scoria (oxidised red and purple) and short, thin flows of lava were also erupted during a later, quieter phase from the largest vent, to build a scoria cone.

These vents were last active just prior to 26,000 years ago. The 26,000-year old Oruanui tephra layer which lies on top of ejecta from the craters provides this date.

Two adjoining explosion craters of a similar age, now filled with water, lie further to the south on the outskirts of Ohakune. The Rangataua Lakes are the southernmost vents of the Taupo Volcanic Zone, the band of volcanic activity which stretches about 240km across the central North Island offshore to the volcano of White Island.

The Ohakune Craters, situated 18km to the south-west of Mount Ruapehu, are parasitic vents of this massif. They are the southernmost vents of the Taupo Volcanic Zone, the band of volcanic activity which stretches about 240km across the central North Island offshore to the volcano of White Island.

Columnar joints sometimes form in cooling lava flows. As the lava cools, it contracts causing cracks or joints to develop perpendicular to the surface. Erosion may later reveal steep, multi-sided columns formed by these cracks. The columns may be 10cm to well over 2m in diameter and tens of metres high. One of the best developed examples of jointing can be seen at "Column Rocks" at about 1700m on the south-west slopes of Ruapehu. These columns average about 15cm across and are exposed in a cliff about 10m high.

RANGATAUA LAVA FLOWS

A vent has been active on the southern slopes of Ruapehu below Girdlestone Peak at an elevation of 1720m (vent area enclosed by box). It lies on the north-east trending line of volcanic activity that is recognised through the massifs of Ruapehu and Tongariro. A low mound of pyroclastic material is all that distinguishes the vent region; however, this is the source of the lava field which now forms the southern shoulder of Ruapehu. Composed of overlapping tongues of block lava, the Rangataua flows are up to 14km long and 4km wide. The flows are at least 150m thick in places and have a volume of about 1.46 cubic kilometres.

The upper eastern margin of the flows is bounded by an Ice Age moraine which indicates that the flows were erupted since the last glaciation. In addition, the lava shows more surface weathering than the Whakapapa flows (on the north-west side of the mountain) which are thought to be younger than 9700 years. Thus, the Rangataua flows have an estimated age of between 10,000 and 15,000 years.

Soil formation has taken place in ash layers overlying the lava, and below a height of 1200m the flows are covered in mature red and silver beech forest.

THERMAL ACTIVITY AND COLD SPRINGS

We *"stood in the centre of a region where boiling springs burst from the earth, where jets of steam shrieked and hissed from innumerable fissures, where enormous boiling mud-holes bubbled like heated cauldrons and where the hot, steaming soil . . . quaked beneath our feet . . . so that we had to pick our way cautiously amid clouds of steam and sulphurous fumes."*

James Kerry-Nicholls, 1884

It is no coincidence that most of New Zealand's thermal activity is located in the centre of the North Island. Hydrothermal features (hot springs, geysers, mud pools and fumaroles) have a natural heat source — magma which tends to lie closer to the surface in volcanic zones, increasing the likelihood of thermal activity.

A great deal of magma never reaches the surface to be erupted from volcanoes. It can remain hot for thousands of years, even after volcanic activity has ceased. Rising hot gases, from slowly cooling magma, heat up the surrounding rocks. Ground water moving down towards the magma is heated through contact with these rocks. It becomes progressively hotter and may be flashed to steam. As hot water is less dense than cold water, it is able to rise to the surface along fault lines where it appears as hot springs, geysers and fumaroles.

Three small thermal areas are found in the Park, at Ketetahi Springs and Te Maari and Red Craters, all on the Tongariro massif. They are believed to share the same underground reservoir of high temperature (200-250°C) steam.

Steaming ground, Ketetahi Springs area.

KETETAHI SPRINGS

A major area of thermal activity occurs at about 1400m on the northern slopes of North Crater, Tongariro. Activity is concentrated within a 30-hectare area along branches of the Mangatipua Stream - the enchanted stream. The Ketetahi Springs are located on a small enclave of private land that is retained by the Ngati Tuwharetoa people and is surrounded by Tongariro National Park. Although the track crosses close to the base of the springs please respect the wishes of the Maori owners and do not visit the thermal area or bathe in the stream.

Vents emitting volcanic gases are called fumaroles which is the Latin name 'to smoke'. The fumaroles of Tongariro vary from small, gently steaming vents to noisy, high pressure steam vents. 'Main Blowhole', the largest fumarole in the Ketetahi Springs area, discharges steam superheated to a temperature of about 138°C at a rate of up to 90m per second.

At left: The track over Tongariro (the Tongariro Crossing) cuts across the flanks of North Crater before zigzagging down to Ketetahi Hut. The Ketetahi Springs lie beyond, a few minutes walk from the hut.

HYDROTHERMAL DEPOSITS AND ALTERATION OF ROCK

The thermal areas of Tongariro, particularly Ketetahi Springs, are brightly coloured. The hot, acidic spring waters and steam are responsible for the myriad of colour ranging from shades of orange, brown, purple and red to white, green and black.

High temperature steam and boiling water can dissolve rocks much more efficiently than cool rainwater. Thus, the hot ground water becomes enriched by a variety of dissolved minerals from the andesite rock the solution moves through. On reaching the surface, the hot water or steam rapidly cools and most of the minerals are precipitated.

The waters of the thermal areas of Tongariro contain boric acid, sulphate, plus lesser amounts of magnesium, calcium, iron and ammonia. These are slowly being deposited around vents and springs.

Hot water and steam can also alter the composition of the andesite rock to form soft multi-coloured clay soils. These range in colour according to the amount and type of minerals present. For example, iron compounds are responsible for a range of red, brown, pink and golden colours, and iron sulphide turns clay and water a blackish colour.

Yellow sulphur deposits are extensive in and near areas of steaming ground because gases emitted from fumaroles are often rich in either sulphur vapour or hydrogen sulphide. Crystals of sulphur form around the vents as the gases cool and oxidise in the atmosphere.

Some hot springs form bubbling pools of coloured mud. Steam and gas escaping from boiling water beneath clay soils produce these hot mud pools.

Colourful red, and blue-green algae grow in the hot acid-sulphate waters of Ketetahi, adding to the spectrum of colour.

Ketetahi Springs — a kaleidoscope of natural colours.

COLD SPRINGS OF THE PARK

Numerous cold springs occur on the slopes of Tongariro and Ruapehu. Rain water is the source of all these springs, yet the chemistry of the spring waters reflects the composition of andesite rock from the volcanoes. Most of the springs are acidic, with a pH range of 4.6 — 6.1. The waters contain high levels of silicates and other minerals in varying amounts. In some cases, the concentration of certain minerals is sufficient to precipitate onto the stream bed. Rocks of the Whakapapanui Stream, for example, are coated with a deposit of an aluminium-rich clay that is leached from andesite ash and is coloured by iron oxide.

SODA SPRINGS

These cold water springs seep to the surface at the head of the Mangatepopo Stream at the north-west base of Ngauruhoe. The water comes to the surface in a boggy area. The rocks at and below the springs are coloured golden by iron oxide which has resulted from the breakdown of volcanic ash in the bog. The water is slightly charged with dissolved gases and this effervescent quality inspired the name Soda Springs.

OHINEPANGO SPRINGS

The Ohinepango Springs gush from beneath an old lava flow to the east of Ngauruhoe, about 1km from Waihohonu Hut. The springs discharge an enormous volume of glassy-clear water into a bush-encircled pool at the head of the Ohinepango Stream.

SILICA SPRINGS AND RAPIDS

An unusual creamy-white deposit coats the bed of the Waikare Stream. It is particularly noticeable through a fast flowing section known as the Silica Rapids. The deposit is derived from mineral-rich spring waters which issue from the base of a lava flow at the head of the valley. The water has probably travelled a long way underground because it is rich in aluminium and silicate minerals from the chemical breakdown of andesite rock. The spring water is also rich in carbon dioxide gas from a geothermal source.

There are no deposits until 50m below the spring, but as the stream gathers speed a thin deposit begins to appear. The deposit is at its thickest (about 3cm) in the turbulent stretches of the stream, especially on the downstream faces of boulders. Here, carbon dioxide is lost quickly from the water, producing ideal conditions for the deposition of the alumino-silicate.

THE LANDSCAPE IS MODIFIED

Successive episodes of volcanic activity during the past one million years have constructed the volcanoes of the Tongariro Centre. Yet the landscape we see today was not created by purely volcanic forces. Modified by rain, running water, snow, ice and wind, it is the end result of both constructive and destructive processes.

In contrast to volcanic activity which can alter the landscape overnight, the processes of erosion generally produce more subtle, sometimes imperceptible changes. However, these continuously active processes inevitably transfigure the landscape with the passage of time. The constructive and destructive forces at work on the volcanoes will continue their antagonist roles, interacting to produce the landscapes of the future.

Mahuia Rapids in flood.

The Whakapapaiti Stream descends, via numerous waterfalls over lava bluffs, from its headwaters high on the north-west slopes of Ruapehu.

"Here you may see the heart bared, see the
process of the making and moulding of the land by
fire, ice and water. Mother-Earth reveals her
inmost secrets here; she pulses with never ceasing,
sometimes fiery, energy."
James Cowan, 1927

EROSION BY WATER

Volcanoes are extremely vulnerable to erosion. Their mantle of loose volcanic ash and rock fragments, often unprotected by vegetation, is easily washed down the steep slopes. Water courses develop, and long term down-cutting establishes a pattern of deep gullies radiating from the mountain tops like the spokes of a wheel.

The massifs of Tongariro and Ruapehu are deeply eroded. Radial drainage feeds three major river systems — the Tongariro, Wanganui and Whangaehu Rivers and numerous other streams. In contrast Mount Ngauruhoe is little dissected. No deep gorges mar its symmetrical shape but a pattern of radial drainage is established on the cone. This may be the forerunner of more intense erosion.

The south-eastern flank is the most eroded side of Ngauruhoe. In contrast to the rest of the volcano no lava has flowed down these slopes for probably hundreds of years — sufficient time for small valleys to be eroded by freeze thaw and stream action. In an earlier cycle of erosion a glacier gouged out this valley down which the Waihohonu Stream now flows.

The processes of frost heave and freeze-thaw action are active on the upper slopes of the volcanoes for much of the year. As the ground freezes, growing ice crystals heave stones and smaller particles of rock up above the surface. When the ice melts, these particles are dumped loosely back onto the ground. Elsewhere, water finds its way into narrow fractures and pores in rocks. Freezing causes expanding ice to exert pressure on the rock, eventually breaking it. In these ways frost action continually loosens surface rock material which is then susceptible to erosion by water.

A slowly changing pattern of relief, in which the valleys of one era become the ridges of the next, often occurs on active volcanic cones.

"Topographic reversal" begins with the eruption of lava into a radial valley. If the valley is filled by lava flows then water runoff from rain or melting snow will be forced to flow beside the lava on the site of former ridges. After many thousands of years of erosion (by streams or glaciers), the ridges will be converted to valleys and the lava flows into free-standing ridges.

This process has occurred time and again on the slopes of Tongariro and Ruapehu, and consequently some of the prominent lava ridges and bluffs of the present landscape are the oldest parts of the massifs.

Cathedral Rocks on the eastern side of Ruapehu illustrate topographic reversal. The series of lava flows which form this ridge were erupted about 200 000 years ago. They drained into and filled an ancient valley, probably displacing a stream or glacier. A new pattern of drainage was established along the margins of the flows and erosion over a long period cut deep valleys on either side. The glaciers of the Whangaehu and the Mangatoetoenui are at least partly responsible for the down-cutting that has isolated the flows and converted them into a ridge.

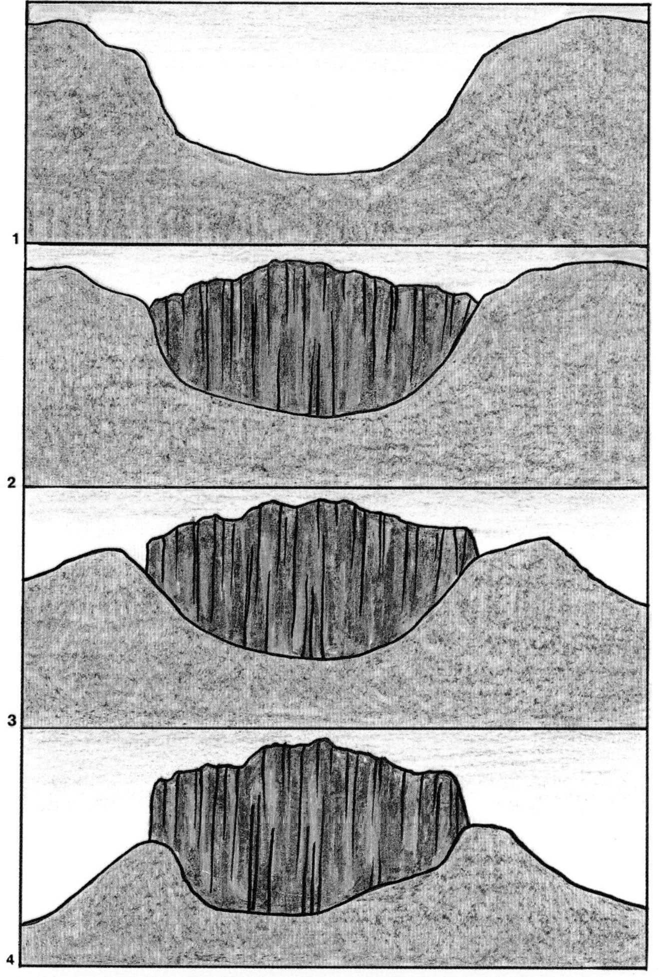

Diagram showing the process of topographic reversal
1. Valley
2. A lava flow fills the valley
3. New valleys begin to develop on either side
4. The lava flow eventually becomes a ridge

LAHARS RESHAPE THE LAND

Lahars have altered the character of the landscape by moving enormous amounts of debris down the slopes of the volcanoes and dumping it at lower levels.

Lahars occurred frequently throughout the last Ice Age. Giant flows, lubricated by great volumes of ice and snow, swept down the mountain sides depositing thick fans of debris. Mudflow deposits, from scores of lahars, overlapped to form broad rings around the base of the volcanoes. Deposits from these Ice Age lahars extend up to 50km in the north-west near Taumarunui and 45km south-east to near Taihape. Boulders of andesite found in the Rangitikei Valley may have been carried up to 75km by lahars from Ruapehu.

Only remnants of the ring plain which surrounded the Kakaramea-Tihia complex survive today. However, the 10-20km wide plains ringing Tongariro and Ruapehu are still largely intact.

Mudflows carry with them a mass of material ranging from house-sized blocks to tiny particles of volcanic ash. The jumbled deposits of lahars can be seen in road cuttings near the volcanoes, for instance, on the road to the Whakapapa Village and in the Makatote and Manganuioteao gorges (100m thick exposures) to the west of Ruapehu.

LAHAR MOUNDS?

More than 100 small mounds extend down the lower north-western slopes of Ruapehu (left), covering an area of about 25 square kilometres on the ring plain near the junction of State Highways 47 and 48. Similar groups of conical hills are found in association with other volcanoes in New Zealand and overseas.

Several theories have been put forward to explain the formation of these mounds: that they are bubbles in lava flows; individual volcanic vents; or glacial moraine deposits. These ideas have been discounted, and it is now widely believed that the mounds are formed in avalanches of coarse rock material. Debris avalanches are dry and poorly mixed in comparison to true lahars, which tend to be wet and muddy. Steep volcanic flanks often become unstable as a result of loading by ejecta or lava flows, over-steepening of slopes during erosion, and saturation due to heavy rain or snow melt. Large blocks of the original mountain can be transported intact for great distances.

In 1914, the inner crater wall of White Island became unstable, and coarse rock material flowed quickly over the crater floor. This debris avalanche formed mounds that are similar in size and appearance to those near Ruapehu.

Research has shown that many of the Ruapehu "lahar mounds" are pieces of the old Te Herenga cone, implying that the debris avalanche originated in the headwaters of the Whakapapanui Stream, probably from the area we now call Pinnacle Ridge. Huge pieces of the mountain were carried downstream in a slurry of smaller rocks, sand and mud. The blocks came to rest and formed a broad fan extending beyond the mounds. This fan can be seen in road cuttings along State Highway 47 on the north side of the Whakapapaiti Stream Valley. It contains many abraded tree limbs that were uprooted and carried in the avalanche.

A number of rounded hillocks known as lahar mounds border the road to Whakapapa Village.

THE IMPACT OF GLACIAL EROSION

Glacial action has made its mark on the volcanic landforms of the Park. The massifs of Tongariro and Ruapehu were extensively eroded by Ice Age glaciers.

Although much of the early record of glaciation has been obliterated by subsequent volcanic activity and water erosion, some characteristic examples of glacial landforms survive in the Park.

We know that Ice Age glaciers occupied the valleys of the Mangahouhounui, Oturere, Waihohonu, Makahikatoa and Mangatepopo on Mount Tongariro — for these valleys have been converted from what were probably narrow stream valleys into broad glacial troughs.

The Mangatepopo Valley is a distinctive example of a U-shaped valley. This landform and the regular ridges of debris (moraines) which line the valley are evidence that the "Mangatepopo Glacier" once extended for about 5km. The tephra cover preserved on these moraines dates the maximum advance of this glacier to just prior to 14 700 years ago.

The floor of the Mangatepopo Valley is gradually being covered with lava flows from Mount Ngauruhoe. Volcanic activity may eventually obscure the remains of the glacial landscape in this area.

Mount Ngauruhoe, Pukekaikiore and the Mangatepopo Valley.

Two views of the Waihohonu Valley moraines — one looking up-valley and the other down-valley.

MORAINES

Rock material that is caught up in a glacier and transported by it gets deposited from the glacier at some stage. These accumulations of rock debris are called moraines. They are commonly deposited along the sides (lateral moraines) or at the end of a glacier (terminal moraines). Lateral moraines, of varying ages, are the best preserved form of moraine in the Park. The prominent pairs of moraine ridges which bound the Mangatepopo, Makahikatoa and Waihohonu Valleys were deposited during the last Ice Age, more than 14 000 years ago. Few moraines of comparable age are as distinctive on Ruapehu, although Ice Age moraines are found on the eastern slopes of the mountain. A curving ridge, low on the north-west slopes of Ruapehu east of the Whakapapaiti Stream, may also be an Ice Age moraine. Several very young lateral moraines are found in the upper Mangaehuehu, Wahianoa and Mangatoetoenui Valleys. These moraines were probably added to as recently as late last century, during a period of minor glacial advance.

GLACIERS — WHAT GLACIERS?

THE PRESENT DAY GLACIERS OF RUAPEHU

Although the glaciers which covered Mount Tongariro during the last Ice Age have long since vanished, a few remain on the upper slopes of Mount Ruapehu. These small glaciers, the longest of which is only 1.8km, are the northernmost glaciers in New Zealand. Covering about 4 square kilometres, or only 0.5% of Tongariro National Park, they are often overlooked in a landscape renowned for its volcanic activity. However, the glaciers are features of the present landscape and they are continuing to have an erosive impact on the shape of Ruapehu.

Most of the surviving glaciers of Ruapehu are "cirque" glaciers. They originate from snowfall at the head of bowl-shaped basins. The Mangaehuehu Glacier is a cirque glacier and the Mangaturuturu, Wahianoa and Crater Basin Glaciers also lie in basins. These glaciers are continuing to excavate and deepen the valleys they occupy.

The Whangaehu, Mangatoetoenui (Waikato) and Whakapapa Glaciers also occupy cirque basins but they receive ice from the Summit Plateau ice field as well. Downwasting of this ice field over the last 30 years has gradually lessened the flow of ice out of it so that the Whangaehu Glacier is now the only true "outlet" glacier. Most of the glaciers of Ruapehu have been shrinking in recent decades, as have many glaciers elsewhere in the world.

In the next few decades further downwasting of ice on Ruapehu may affect the stability of rock slopes and cliffs beside the glaciers, such as south-west of Pyramid Peak.

Late summer, and a crevasse yawns wide in Crater Basin Glacier beneath the summit peak of Ruapehu.

"Ruapehu, 9,200ft. in altitude, has glaciers on its east, south and west slopes, which, although they cannot vie with their southern sisters in magnitude or beauty, are the only ice-rivers of the North Island."

Leonard Cockayne, 1908

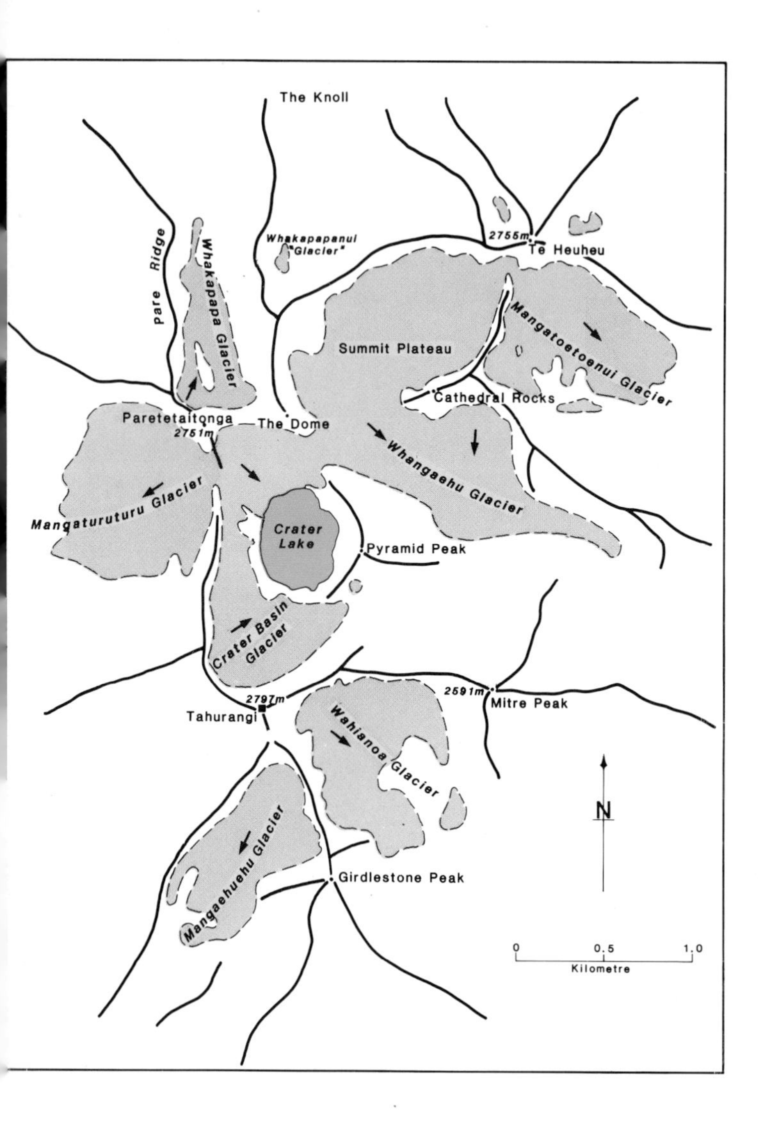

This map shows the location and extent of the present day glaciers on Ruapehu. It is based on a survey made in autumn 1988. Arrows show the direction of ice flow.

GLACIAL FLOW RATES

Despite their small size, the glaciers of Ruapehu are true glaciers. They erode bedrock and form moraines like their Ice Age counterparts, but on a reduced scale. To do this a glacier must flow. Glacier ice is not a brittle solid. It moves slowly by internal flowage downhill. It is difficult to measure accurately how fast a glacier moves. Measurements have to be made over at least several years as glacier flow is often very slight over a short period.

In 1951, wreckage of a plane became incorporated into the Mangatoetoenui Glacier on the eastern side of Ruapehu, at an elevation of 2600m. Thirty-one years later the plane emerged, about 600m further down the glacier. The wreckage was pinpointed again in March 1988 and its position then made it possible to calculate that the Mangatoetoenui Glacier is moving at an average speed of about 20m per year. This rate, although relatively slow (the Tasman Glacier at Mount Cook moves about 100-200m per year), indicates that the glacier is still active. Similarly, lines of dirt ridges on ice draining from the Summit ice field into the Whangaehu Glacier are moved at an average speed of 7m per year.

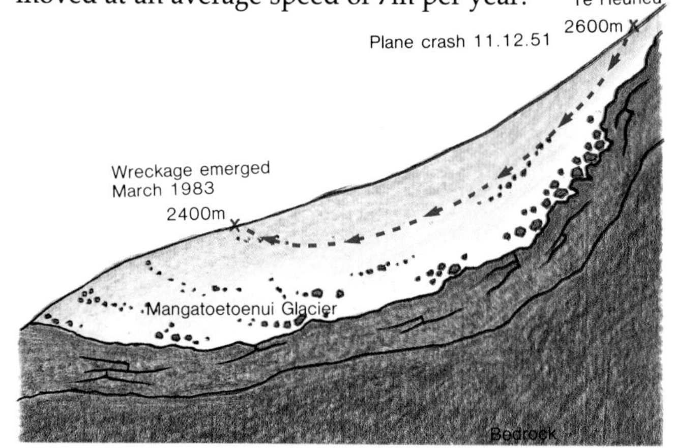

CRATER BASIN GLACIER

Glacier ice extends to the north, west and south of Crater Lake. The ice cover in this area has changed substantially since the 1950s.

An 8m drop in lake level following the disastrous Tangiwai lahar left ice cliffs bordering the north side of the lake high and dry about 200m back from the lake edge. After 1953, glacier ice accumulated and advanced to fill the gap, and new cliffs bordered the northern end of the lake up until the 1995 - 96 eruptions. During this period the ice continued to thin in the rest of the Crater Basin, by as much as 90m in the south below Tahurangi.

This different behaviour and thinning have revealed that the crater basin in fact contains two glaciers. The Tuwharetoa Glacier which flows off Paretetaitonga to the Crater Lake is the most active and "healthy" of all Ruapehu's glaciers. The other glacier in the Crater Basin is situated under Te Ata Ahua and Tahurangi. This southern glacier is presently melting at a rate of 3-5m a year.

Compare these two photographs of Crater Basin Glacier taken 64 years apart. The black and white photo of the glacier in April 1909 shows an ice cliff more than 60m high extending around the south side of Crater Lake.

Considerable thinning and glacial retreat has taken place in the area surrounding Crater Lake as is apparent in this January 1973 photo - the area beneath the summit of Ruapehu is now concave and the icecliffs seen in the photo above are gone. The outlet of Crater Lake can be seen directly below the summit. Up until 1960 lake water discharged into a cave in glacier ice by the lake shore. By the 1990s the glacier had retreated so far back that lake water flowed over a series of waterfalls before entering an ice cave that was 120m from the lake and 40m below lake level. This ice cave was destroyed in the eruptions of September 1995 and other changes occurred.

MANGAEHUEHU GLACIER

In the early 1900s the Mangaehuehu Glacier, on the south-west slopes of Ruapehu, extended almost 2km down the Mangaehuehu Valley where it terminated in a 15m ice face. This is surveyor Hubert Girdlestone's description of the glacier in 1909:

"In March this glacier presents a magnificent spectacle, being cracked from top to bottom with large crevasses, and about halfway down there is a remarkable icefall which presents a series of tremendous crevasses, ice cliffs, razor-backed ice pinnacles and dark caverns, the greenish tint of the ice giving the place a beautiful effect."

An old postcard of the Mangaehuehu Glacier at the turn of the century. The distinctive peak overlooking the glacier has been sculptured into a feature called a horn which is a characteristic feature of glaciated landscapes. This triangular-shaped peak was originally called the Little, or Lesser Matterhorn by Hubert Girdlestone but after his death in 1918 it was renamed Girdlestone Peak.

The Mangaehuehu Glacier in the 1970s showing the dramatic recession which has taken place this century.

The steep lateral moraine ridges now rise up to 100m above the present glacier margins and can be traced for more than 1km down-valley. These features of glacial deposition can be seen from Wanganui, 90km away.

The Mangaehuehu Glacier has retreated 800-900m up-valley since 1900 and now ends at the foot of a bluff at an elevation of 2130m. This photo taken in June 1984 above the bluff shows a small crevassed area which is probably the remnant of Girdlestone's icefall.

DEATH OF A GLACIER

Although Ruapehu's glaciers have been in mild decline for some time, the first major retreat was witnessed in the mid-1950s. Higher temperatures combined with several dry winters, led to accelerated melting and ice margin retreat. This was particularly noticeable on the Whakapapa Glacier on the north-western slopes of the mountain.

In 1954, the Whakapapa Glacier was about 1.7km long and extended as far as The Knoll (2244m). However, by the end of 1955 the glacier had shrunk nearly 100m in length and thinned about 10m overall. In the following year, the Whakapapa Glacier retreated a further 65m.

At the end of the summer of 1956, a rocky ridge (Restful Rocks) had emerged from beneath the glacier and divided it into two distinct lobes — the Whakapapanui and the Whakapapaiti Glaciers. The north-east branch, the Whakapapanui, was now completely cut off from its sources of supply, the Summit Plateau and the snowfields of the Whakapapaiti Glacier. By the 1960s, the Whakapapanui Glacier was only 10m thick.

The final melting of the Whakapapanui took place in the 1970s. In the summer of 1970 the glacier retreated more than 130m in two months and was reduced to pockets of glacier ice, separated by huge cave-like meltwater holes. Further melting since has rendered the Whakapapanui a glacier in name only, although some snow patches in the area last through the summer.

Glaciers are extremely sensitive indicators of climate change and their equilibrium is easily upset. The glaciers of Ruapehu may have melted completely and subsequently reformed during the life of this volcano. The fluctuating glaciers are another example of the Park's dynamic landscape.

The Whakapapanui Glacier, January 1955, during a period of particularly rapid retreat. Mr L.O. Krenek of Wanganui wrote at the time: "The glaciers of Ruapehu are simply collapsing. The Whakapapa has receded 75 yards (68m) since last year, and enormous areas of rock have emerged."

Rock fragments picked up by a glacier may become enclosed in glacial ice. When this debris is dragged across the ground beneath the moving glacier, it can act like sandpaper — smoothing and polishing underlying rocks, or grooving and scratching the valley floor.

The recent retreat of the Whakapapa Glacier has exposed bedrock surfaces that have been shaped and polished by overriding ice. Glacial scratches (striations), showing the direction of glacier motion, can also be found on rock surfaces in the area. Subsequent weathering removes most of these scratches within about a hundred years.

THE VOLCANIC LABORATORY

The active volcanoes of the Tongariro Volcanic Centre are a natural laboratory for research. They are accessible and even when an eruption is in progress they can be viewed from close by in relative safety.

Volcanic activity is monitored by the Volcanic Research Group of the Institute of Geological and Nuclear Sciences (IGNS). This organisation undertakes geophysical, deformational and chemical studies throughout the Taupo Volcanic Zone. A number of universities are also involved with fieldwork and volcanological studies in the area.

Seismic activity, relative changes in the Earth's magnetic field and atmospheric shockwaves are recorded automatically at the IGNS observatory in Whakapapa village.

Study of Ruapehu and Ngauruhoe intensifies during active periods as much can be learned at these times. Scientists are working to identify eruption patterns, frequency and volume and to gather other information which will help in the mitigation of volcanic risk. They are also seeking to explain the diversity in composition of eruptive products and reconcile this with what is happening in the shallow 'plumbing systems' beneath these volcanoes. A summary of volcanic observations is published annually in the NZ Volcanological Record.

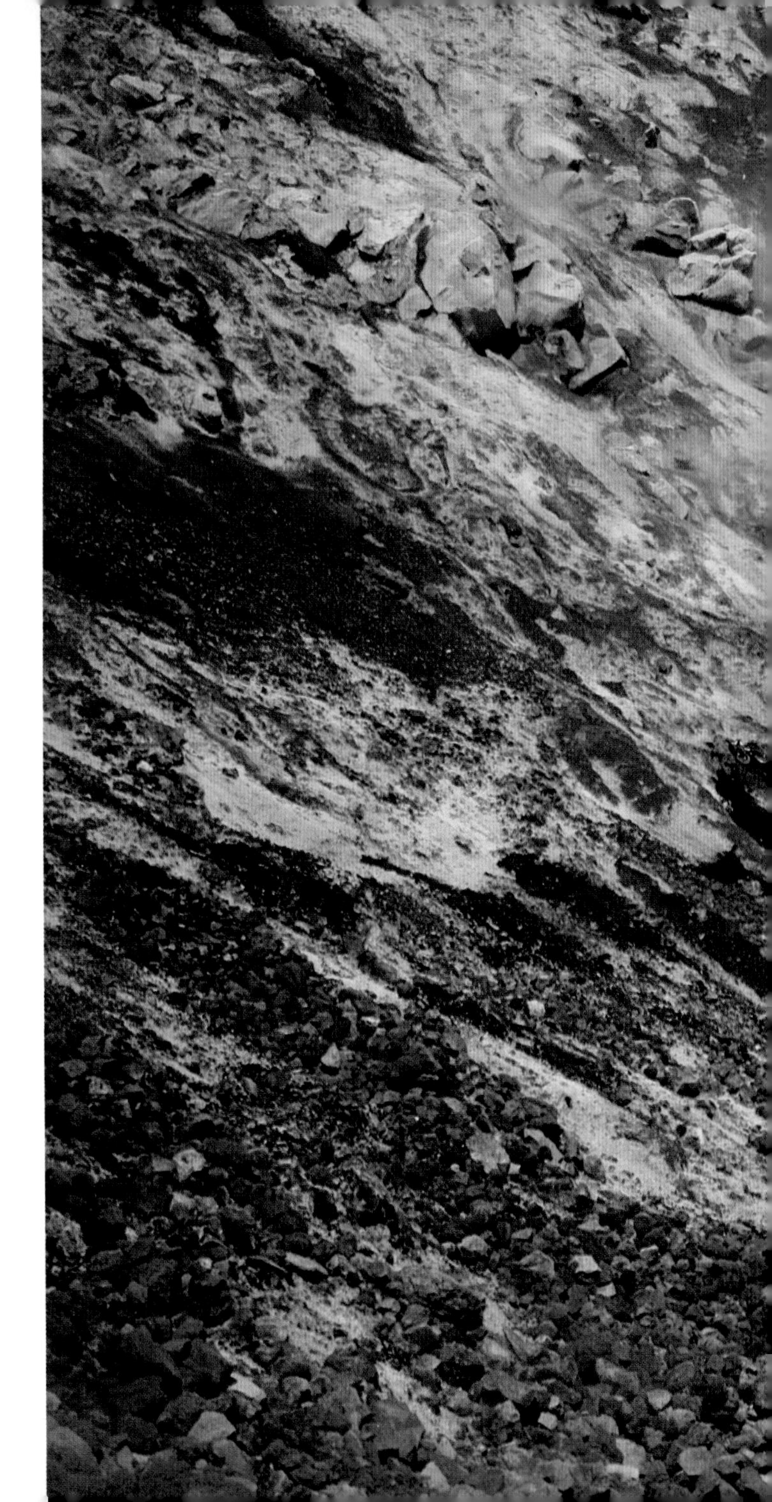

A scientist samples hot volcanic gas in the bottom of the crater of Ngauruhoe.

"We should be doing all we can to understand our volcanoes, as New Zealand is one of the finest areas in the world for their study. What can be learned here is not merely of local interest, but is vitally useful to other countries."
Colin Wilson, 1982

MAGNETIC SURVEYS

Volcanoes have characteristic magnetic fields, and the magnetism of volcanic rocks decreases with increasing temperature. Therefore, it should be possible to measure changes in the magnetic strength of a volcano before or during an eruption. Magnetic field measurements have been carried out on Ruapehu in recent years, using instruments called magnetometers. So far, no variations in magnetism that can be attributed to volcanic activity have been detected.

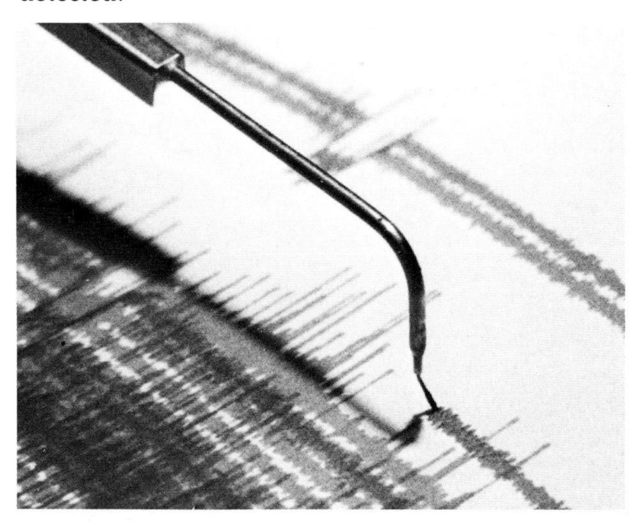

Seismographs produce a continuous record of minute vibrations which may be caused by volcanic activity or earthquakes. Analyses of seismic records from Tongariro National Park are adding to information about eruptive patterns.

SEISMIC STUDIES

Seismometers, the basic instrument used in the study of earth vibrations, continuously record seismic activity in the Park. Volcanic tremor can be used to provide early warning of increased volcanic activity. Eruptions of Ngauruhoe and Ruapehu are often preceded by tremor (sometimes by up to two weeks) which become more intense as an eruption approaches. However, tremor only gave 29 minutes warning of the destructive 22 June 1969 eruption of Ruapehu, and nine minutes for the eruption of 24 April 1975.

DEFORMATION SURVEYS

Scientists know that the topography of volcanoes can change prior to eruptions. For example, rising magma and gas pressure displaced the flank of Mount St Helens by an impressive 100m before the 1980 eruptions. Similar ground movements, but on a much smaller scale, have helped predict eruptions of White Island.

Precise measurements, between several fixed survey points around the rim of Ruapehu's Crater Lake, are made on a monthly basis. In this way changes, which are imperceptible to the human eye, can be noted.

In addition, precise levelling surveys are used to measure any tilting of the Earth's surface in the vicinity of Ngauruhoe, Ruapehu and Tongariro. The results have been encouraging. They show a positive correlation between earth deformation and volcanic activity.

CHEMICAL INVESTIGATIONS

Ruapehu's Crater Lake is inspected routinely by staff from the Institute of Geological and Nuclear Sciences at Wairakei. Observations include lake and air temperatures, lake level and overflow rate, and evidence of recent eruptions or convection within the lake. Water samples are also collected for analysis.

The temperature and composition of fumaroles within the crater of Ngauruhoe, Red Crater and Ketetahi Springs are also monitored by IGNS scientists. The acidity of the Whangaehu River, into which Crater Lake overflows, is measured regularly by Tranz Rail and the Rangitikei-Wanganui Catchment Board at Tangiwai.

The chemistry of the water in Crater Lake is complex and variable. Water samples have been collected since the 1960s and have revealed a correlation between chemical and volcanic activity.

The pH of the lake water is an important chemical indicator associated with eruptive activity. The fluctuating concentrations and relative proportion of magnesium and chloride (two major constituents of Crater Lake water) are also reliable indicators of whether the processes of degassing and the extrusion of magma are occurring. Analyses suggest that the interaction of hot andesitic rock (magmatic material) with lake water increases magnesium concentrations, whereas the amount of chloride in the lake is related to fumarolic activity.

Samples taken 8-15 months prior to major eruptions in 1968, 1969, 1971 and 1975 showed constant magnesium-chloride ratios. Such ratios are typical of blocked vent conditions and suggest an increased probability of eruptions. However, more data are required before eruptions can be predicted in this way.

WHAT'S UNDERNEATH CRATER LAKE?

Unlike Mount Ngauruhoe, the floor of Ruapehu's active crater is concealed by a lake. Because changes in lake depths can reflect changes occurring in the active vent, scientists measure the depth of Crater Lake from time to time.

The floor of the crater has only been observed during 1945-46 when the lake was temporarily displaced by lava. A central vent with a depth of 300m was noted before the lake began to refill.

Depth soundings made in 1965 showed there had been little change in the lake floor since 1946. The survey confirmed the continued existence of a 300m deep, funnel-shaped central vent. However, a survey in 1970 revealed the lake to be significantly shallower, ranging between 50m and 75m deep. Infilling probably took place when magma was injected into Crater Lake during an active period in 1966. The eruptions of 1968-69 may also have contributed to the lake bottom changes.

During an active phase in 1982, scientists used remote depth-sounding buoys to discover whether lava was being extruded onto the crater floor. They discovered the depth of Crater Lake exceeded 180m in a cone-shaped central area, and that there had been no major extrusion of lava.

A detailed survey of the depth of Ruapehu's summit lake has only been carried out three times. This potentially hazardous task was accomplished in 1965 and 1970 from small boats using depth-and echo-sounding equipment. During an eruptive period in 1982, 10 depth-sounding buoys were dropped into position on the lake by helicopter.

VOLCANIC HISTORY

Information about the eruptive history of the composite volcanoes of the Tongariro Volcanic Centre continues to grow based on detailed geological mapping of volcanic stratigraphy and tephra layers, sampling and the application of chemical dating techniques.

The volcanic history of the past 20 000 years or so is relatively accessible and best understood but the source and type of prehistoric eruptions are of great interest. They reveal how often certain types of events occurred in the past and give some indication of what could happen in the future.

Radiometric techniques are used to determine the age of volcanic rocks or tephra deposits by comparing the amount of a particular radioactive substance they contain and the amount of its decay products. Best known is the carbon-14 technique, however, material older than about 40 000 years cannot be dated by this method. K-Ar dating (based on the ratio of potassium to argon), is used to date rocks older than 100 000 years. Both dating methods have been used extensively in the Park, adding to the geochemical data base.

Scientific advancements in this area will continue to contribute to a more comprehensive reconstruction of the magmatic history of the Tongariro Volcanic Centre.

A group studies a series of distinctive tephra layers in Tongariro National Park.

RUAPEHU 1995/1996

A series of spectacular and powerful eruptions rocked Mt Ruapehu for several weeks during both 1995 and 1996. This contrasted with eruptions during the previous five decades which had generally been of short duration and included single, large, 'one shot' eruptions such as in 1969 and 1975.

Wide variations in Crater Lake temperature had been measured from 1991 at Ruapehu. In late 1994 water temperatures climbed from a low of 15°C in October to reach 51°C in January 1995. In May 1995, a major change in lake water chemistry indicated fresh magma was influencing the volcano system. When this was followed by a period of intense volcanic tremor in late June scientists began to take a close interest in what was happening at Ruapehu.

An eruption generating lahars down into the Whangaehu catchment on 29 June destroyed volcano monitoring equipment on the lakeshore. This event was followed by a cooling of the lake to 29°C during July-August and varying seismicity.

The main eruption sequence began at 8am on 18 September with a single moderately-sized explosion through Crater Lake. This produced the largest mudflow on Ruapehu since 1975 and destroyed the Whangaehu footbridge on the Round the Mountain Track.

At 5pm on 23 September a sequence of major Surtseyian-style eruptions took place. Volcanic ash, bombs, blocks and hot lake water were violently ejected onto the summit areas around the crater. Turbulent clouds of steam and ash surged outwards from Crater Lake. Volcanic blocks trailing white arcs of steam were tossed above the summit to land up to 1.5km from the crater.

Left: An aerial view of the ongoing eruptions - 29 September 1995.

The largest single eruption of the 1995 sequence occurred at 5pm on 23 September. The explosions carried lake water, ash and huge chunks of lava to hundreds of metres above the summit of the mountain.

MUDFLOWS

The eruptions of 23 September punched through Ruapehu's Crater Lake sending high velocity jets of ash, debris and hot lake water laterally into three main valleys. Most of the lake water was displaced and flooded down the Whangaehu, Whakapapaiti and Mangaturuturu valleys. At its peak the Whangaehu River rose 2.6m up the lahar warning gauge 10km upstream from Tangiwai.

Shallow eruptions continued from Crater Lake into October 1995. More than 30 mudflows were generated, setting a new record in terms of number, frequency and total volume of lahars for the past 120 years. The largest lahar flowed eastward down the Whangaehu River, the natural outlet of Crater Lake. As the eruptions continued, the threat of more dangerous mudflows reduced as the level of the water in the lake steadily diminished.

VIOLENT EXPLOSIONS

Numerous small to moderate explosions, often only 1-5 minutes apart, began on the evening of 24 September. The eruptions continued into the next day with even more intensity until they were less than a minute apart during their peak between noon and 10pm. Violent explosions took place as molten magma rising in the vent encountered lake water and was blown into pieces by high pressure and steam. These explosions generated a plume of ash and gas above the volcano forming a continuous 10km-high eruption column.

Ash fell as far away as Hawke's Bay and forced the temporary closure of the Desert Road. The Tukino Skifield was covered in ash and was closed for the rest of the ski season. Whakapapa and Turoa skifields were closed for nearly three weeks due to the risk of mudflows and rocks falling on their upper slopes.

Lubricated by water, snow and ice, this material produced several lahars, two of which sped down the Whakapapaiti Valley. These were the first volcanic mudflows to be witnessed and filmed at Ruapehu. They raced down the southern boundary of the ski area at 90kph, or a terrifying 15-25m per second!

Continuous eruptions formed a spectacular plume above Ruapehu on 27 September 1995.

THE LAKE DISAPPEARS

A second intensive period of volcanic activity began on 7 October 1995 when an eruption lasting 15 minutes sent an explosion column to 8km above the mountain. Further eruptions on 10 October triggered a small lahar leaving behind only small pools of water in the bottom of the crater.

The largest sustained eruption of the 1995/96 sequence began at 9pm on 11 October 1995. Later that evening, for the first time, rising magma reached the surface in the crater without encountering water. This resulted in a change in eruption style with new eruptions being controlled by explosive gas release. Around midnight, red-hot rocks fountained above the summit and molten ejecta was flung from vents in the dry lakebed.

Short-lived Strombolian style eruptions occurred during both 1995 and 1996 when the Crater Lake Basin was empty of water. This photo was taken on the night of 17 June 1996 when eruptions propelled molten lava in glowing sprays above the crater. One of the most spectacular displays was captured by this time-lapse photo.

There were no lava flows during the 1995/96 eruptions of Ruapehu. The eruption products were andesitic material from the finest ash to larger pieces of scoria and bombs. These pyroclastic fragments were tossed hundreds of metres in the air and strewn about the summit area. An area up to 5km in diameter was designated a zone of 'extreme risk' at the height of the eruptions.

Dome Shelter, only 300m from the lake, was plastered with debris and bombarded by ejecta which punched holes in the roof and walls of the hut. Fortunately, important seismic equipment installed beneath the hut in a purpose-built cellar was not affected.

GAS EMISSION

Although the eruptions of 1995 were over, large quantities of steam and gas continued to be emitted from Ruapehu for months. The gases included sulphur dioxide (SO_2), hydrogen sulphide (H_2S), carbon dioxide (CO_2) and hydrogen chloride (HCl).

Measurements revealed unusually high amounts of sulphur dioxide (2000 to greater than 10 000 tonnes per day). Volcanoes are a major source of natural SO_2 but this was a much greater rate of discharge than measured from recent eruptions of Mt Pinatubo or Mt Etna. Most of it appeared to be produced from old sulphur deposits being heated and oxidised in the crater system. At times, gas concentrations were unpleasantly high within 4km of the crater.

The light-brown sulphuric gas plume was tracked by satellite on 27 October. The vog (volcanic smog), a result of a reaction between SO_2, oxygen, water and sunlight, extended over the North Island and out to sea off the coast of Christchurch. The fog was eventually dissipated by high winds and heavy rain.

For only the second time in 100 years the Crater Lake had been completely emptied by volcanic activity. Above: Ruapehu's Crater Basin, an awesome sight without water in mid-October 1995. With activity over for the time being the volcano began to cool. Water from melting ice and precipitation in the summit catchment began to accumulate in the bottom of the two active vents.

Top right: The filling process was initially quite slow. By 17 February 1997, the lake level remained about 150m below normal lake level and the northern vent was still dry.

Right: The refilling lake in April 1998 about a year and a half after eruptions had ceased. At this stage the lake was about 9 per cent full and 70m below normal level.

VOLCANIC ASH

The eruptions of 11/12 October 1995 produced the largest volcanic ashfall in New Zealand since 1945. A continuous plume rose to a height of about 10km and was blown in an east-north-east direction by strong winds. Between 10 and 20cm of ash were deposited on the eastern side of Ruapehu and lesser amounts further afield. Two days later more ash fell to the south-east and then, for several months, the volcano was quiet.

Ruapehu sprang back into life with very little warning seven months later on 17 June 1996. Magma welled up into the crater producing violent explosions which sent a small lahar down the upper Whangaehu Valley and once again emptied the lake. The explosive expansion of magmatic gases continued to tear the rising magma apart, blasting pulverised fragments into the air and generating a giant plume of ash. Ongoing eruptions deposited a thin coating of ash over many North Island areas.

IMPACTS

In retrospect, the 1995/96 eruptions were disruptive rather than life threatening. Hazards were mostly confined to the summit area of the mountain and there was little danger to local communities from direct volcanic action. However, the eruptions were extremely costly to the ski industry, closed airports, damaged hydroelectric power facilities and had a regional economic impact of around $100 million. Yet these eruptions hardly figure on a world scale. It is estimated that the combined eruptions produced only one-tenth as much tephra as the 1980 Mt St Helen's eruption in the USA.

At left: The spectacular plume of 17 June 1996 towers above sightseers on the Desert Road. The dark ash-rich plume reached a height of 20km as it was blown northward by a strong wind.

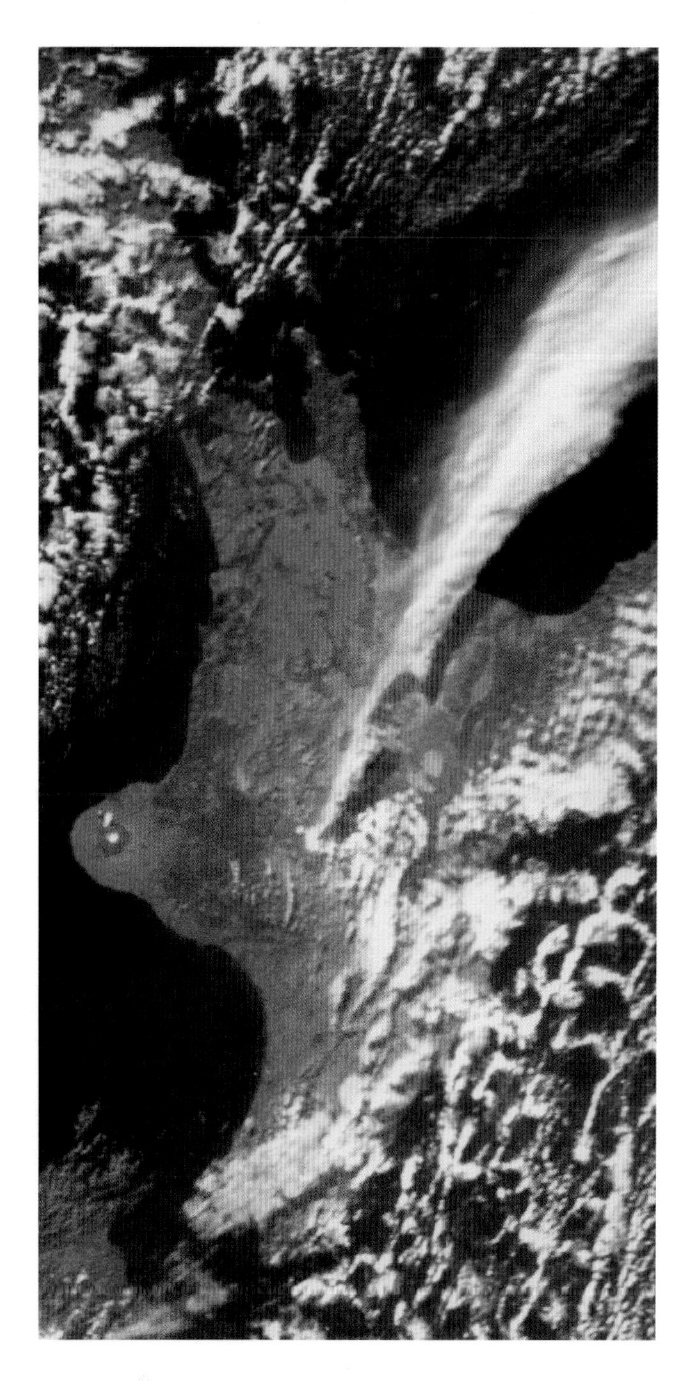

The trail of ash was recorded in this remarkable satellite photo of the North Island.

UNDERSTANDING RUAPEHU

The 1995/96 eruptions provided scientists with new opportunities to understand and explain the processes by which magma (liquid rock) makes its way through a volcano to the surface.

Records show that seismic tremor (near-constant vibration) is common at Ruapehu and that deep volcano-tectonic earthquakes are rare. There is not always adequate precursor activity to warn of eruptions. These factors point to the existence of a shallow active open-vent system with high heat flow.

Scientists suggest the 1995/96 eruptions of Ruapehu were triggered by an influx of new magma into the volcano's shallow 'plumbing system'. Magma, possibly propelled by gas or though a combination of gas and thermally-induced buoyancy,

probably enlarged existing magma pockets or formed new ones until a final pulse made the system unstable. Alternatively, instability may have been created by small pockets of magma coalescing and intermingling.

A consequence of such a shallow magma storage system (less than 1-2km) is that sudden instability, movement or contact between magma and water beneath the lake can cause sudden eruptions with little or no warning. This could explain the so-called 'blue sky' eruptions (e.g., the eruptions of 1969, 1975 and 1988) which are difficult to predict using the usual volcano monitoring techniques.

WHAT HAPPENS WHEN THE LAKE REFILLS?

The eruptions of 1995/96 eroded parts of Ruapehu's crater rim causing serious concerns about what might happen when the lake refills. An investigation concluded that the crater rim was sound but identified a weak dam in the outlet area, built of volcanic deposits, which would probably fail when the Crater Lake eventually refills against it. The geologic report suggested that in a worst case scenario the barrier outlet might suddenly collapse causing a catastrophic flood of up to 1.5 million cubic metres of water at twice the flood discharge rate that caused the Tangiwai disaster.

Subsequently, options to avoid or mitigate the risks posed by a major lahar of this kind were identified. These included installing a lahar warning system for the Whangaehu Valley, restoring the natural outlet by using machinery to dig a channel through the ash blockage, or building a dam and stopbank structures downstream to control the predicted lahar.

However, because the crater lies at the heart of Tongariro National Park, there are cultural, national and world heritage park issues to consider. In May 2000, it was decided not to intervene physically but to install a public alarm system to give warnings of lahars in the Whangaehu Valley.

Crater Lake began to fill during September 1996 and the floor of the whole crater was covered by mid-1997. The lake filled at an average rate of about 900 000 cubic metres per year up to March 2000 but the infilling was not constant. Seasonal rises and falls are controlled by changes in meltwater input and lake water evaporation and it is therefore difficult to estimate when the lake will refill. Estimates range between 2002 and 2008.

RISK MANAGEMENT

The past history of volcanic activity provides the basis for assessing present day volcanic risk at Tongariro National Park. Three main kinds of volcanic hazard have been identified at Ruapehu: lahars (the principal hazard), tephra (ballistic blocks, other projectiles and ash) and volcanic gas. These risks are largely confined to the volcano itself.

The accessibility and attractiveness of Ruapehu's active crater mean that good risk management systems are essential. There has been no significant change in the eruptive styles of the Tongariro volcanoes within historic time but as the recreational use of the park increases, more and more people are at risk from volcanic activity.

Of the more than 30 volcanoes in the world which have skifields on their slopes and have erupted since AD 1800, Ruapehu is among the five most active. Skier fatalities related to volcanic activity have occurred at several overseas volcanoes. Volcanic risk is something which needs to be taken seriously.

A dramatic view of the September 1995 eruptions from the Whakapapa Ski Area.

VOLCANO ALERT SYSTEM

In New Zealand, volcano alert levels are judged and assigned by the Institute of Geological and Nuclear Sciences scientists using monitored aspects of volcanic and seismic activity, observations and measurements. Levels are graded 0 to 5.

The alert system provides a basis for warnings and recommendations of how close the crater area of Ruapehu should be approached. Sensible 'no go' areas of high risk are delineated particularly around the crater during active periods when visitors are warned to stay well clear.

The 1995/96 eruptions provided the first real test of the volcano alert system. The alert level was raised initially from level 1 to level 2. With each significant increase in activity the level was raised to peak at level 4 at the height of the eruptions in September 1995. This signified a hazardous local eruption in progress.

Since December 1997, the alert status at Ruapehu has remained at level 1. Isolated explosive eruptions can still occur at this level.

THE LAHAR RISK

Lahars and their impact-forces destroy most man-made structures and they are the principal hazard to human life at Ruapehu. Ongoing education is essential to ensure people understand they must get out of stream channels and move to higher ground.

Dozens of major lahars have occurred in the Whangaehu Valley since 1861 but only a few have been recorded in the Whakapapa valleys (1895, 1969, 1975 and 1995) during the same period. Yet the Whakapapa area is the principal area of risk because in winter up to 8000 people a day may visit the skifield and some of its valleys are mapped lahar pathways. Turoa Ski Resort is afforded some protection from lahars by the topography of the summit region but it might be threatened by larger lahars and airborne rocks.

Crater Lake Basin, 17 February 1997.

ERUPTION DETECTION SYSTEM

A lahar warning system based on the detection of eruption earthquakes has operated on Ruapehu since 1985, however, the 1995/96 eruptions revealed a need for improvement.

An upgraded alarm system was installed at Ruapehu in 1999. It has the ability to detect eruption air blasts and provide warnings of potential lahars before or as they happen. The new system gives skiers at Whakapapa Ski Area between 30 seconds and five minutes warning and residents in the village a further 15-20 minutes, thus reducing the risk in known lahar pathways.

Tranz Rail's flood gauge on the Whangaehu River upstream of the Tangiwai Bridge continues to give early warning of mudflows and, if necessary, trains can be prevented from crossing the bridge.

Seismicity is continuously monitored at Ruapehu. The seismic signal from The Dome is radio-telemetred to the Chateau Volcano Observatory along with the signal from other seismometers installed at The Chateau, Whakapapa Ski Area, Tukino and near Mt Ngauruhoe. An electronic comparison of these seismic signals can quickly determine whether an earthquake is possibly associated with hazardous volcanic activity.

Lahars look and behave like fast-moving wet concrete as this close-up of a mudflow in the Whangaehu Gorge on the eastern side of Ruapehu in October 1995 shows.

Crater Lake taken from near the outlet on 19 February 2001 - the refilling lake is 47 per cent full and 30m below normal level. The mantle of 1995/96 tephra blocking the outlet can seen on the lower right of this photo.

PARK ROCKS

A CLOSER LOOK

by Dr Bruce Hougton - former IGNS volcanologist, now Chair of Volcanology at the University of Hawaii.

With a little practice anyone can make geological observations. Most people are more familiar with the broader scale of observations of landforms and so here we will concentrate on techniques for describing single rock specimens; like any boulder you might find on a park track or a fist-sized portion of rock broken from an outcrop. We will here limit discussion to volcanic rocks. Observations fall into three categories: colour, texture, and mineralogy (the study of mineral composition, structure and appearance).

Rocks can be studied with the minimum of equipment but useful accessories are:
A handlens or magnifying glass
A pocket knife
A small magnet.

COLOUR: Observations of colour give a clue to rock composition.

White/cream means the rock is probably of *rhyolitic* composition and has its origin in volcanic eruptions from Taupo.

Grey and *black* are typical colours of the *andesite* and *dacite* rocks which make up Tongariro, Ngauruhoe, and Ruapehu. The dark colours mean the rocks are richer in iron and magnesium than rhyolites.

Red generally means that a volcanic rock has been oxidised as it cooled from a molten state. Iron in the rock reacts with atmospheric oxygen to form red iron oxide minerals chemically and physically similar to rust.

TEXTURE: Texture describes the surface features of specimens; the presence or absence of grains, minerals, banding, gas cavities (vesicles) and numerous other features. Some observations are:

Density: Very light rock specimens are generally found to contain numerous spherical or cylinderical *vesicles* which are cavities formed during escape of gas from the lava while it is in a molten state. We call such rocks vesicular and high *vesicular* rhyolites or dacites are known as *pumice* and highly vesicular basalt as *scoria*.

Banding: Banding or layering in volcanic rocks is generally produced during the latter stages of flow of lava as it becomes cooler and less fluid. The banding is generally on a centimetre or millimetre scale and may show spectacular loops and folds.

Crystal Size. Volcanic rocks may consist (1) entirely of large crystals, or (2) contain large crystals (*phenocrysts*) set in volcanic glass, or (3) consist entirely of glass or fine-grained crystals invisible to the naked eye. Crystal size reflects the speed with which the molten rock cools. If lava is trapped beneath a volcano and cools very slowly large crystals can grow and ultimately join together. If lava rises very quickly to the surface it is then chilled quickly by the atmosphere leaving little time for mineral growth and producing a natural glass. We call rocks with large crystals frozen in glass *porphyritic* and they represent lava with a two stage history of cooling — initially slow cooling producing large crystals suspended in liquid then rapid movement to the surface freezing the liquid to glass. Ruapehu and Tongariro lava is typically porphyritic with large white crystals of plagioclase set in glass.

MINERALOGY: Four minerals are commonly visible in rock specimens from the park.

Plagioclase: A milky white sodium and calcium silicate generally up to several millimetres in size.

Pyroxene: A black or dark green iron, magnesium, and calcium silicate which breaks (cleaves) in perfect right angles.

Olivine: A green iron, magnesium silicate without good cleavage.

Magnetite: A black, shiny iron oxide mineral attracted to the magnet.

A number of minerals commonly found in Tongariro Volcanic Centre lavas are visible in this piece of andesite from the Mahuia Rapids lava flows.

GLOSSARY

aa lava
a Hawaiian word for the rough and jagged surface of lava flows which results from the top surface cooling and hardening while the interior of the flow is still moving

acidic
volcanic rocks of more than 67% silica

andesite
a type of volcanic rock, intermediate in composition between rhyolite and basalt

ash
fine volcanic rock fragments, of less than 2mm in diameter, formed by escaping gases pulverising magma. Volcanic ash is not the product of burning

basalt
a type of volcanic rock — a dark coloured lava containing less silica than andesite

blocks
angular pyroclastic fragments (greater than 64mm in diameter) which were ejected as a solid

block lava
flows which cool forming a mass of angular, smooth-sided blocks of solid lava

bombs
fragments of wholly or partly molten lava (greater than 64mm in diameter) which acquire a rounded, or distinctive shape as they travel through the air

breadcrust bomb
a volcanic bomb with a characteristic "crusty" surface formed when trapped gases cause the outer skin to crack during fast solidification and cooling

caldera
a large (greater than 1km wide), more or less circular depression, usually formed by collapse following volcanic eruptions

composite cone
a volcanic cone built of alternating layers of lava and pyroclastic material, sometimes called a strato-volcano

crater
a circular depression, with steep walls, out of which the products of volcanic eruptions are ejected or flow

crust
the outside "skin" of the Earth consisting of a rigid rock layer that is up to 60km thick

dacite
a type of volcanic rock which lies between rhyolite and andesite in chemical composition.

dike (dyke)
a body of magma which has intruded into and cut across the structure of the host rock, and on cooling become a vertical sheet of rock

explosion crater
a crater formed chiefly, or entirely, by explosion

fault
a surface or zone of rock fracture along which there has been displacement, that is, movement along each side of the fault, relative to the other

fumarole
a small volcanic vent emitting volcanic gases and vapours

graben
a down-dropped block bordered by approximately parallel fault scarps; a tensional feature

greywacke
grey, hard sedimentary rock originally deposited in the sea as layers of mud and sand

hydrothermal activity
any process involving high temperature ground water, especially the alteration and deposition of minerals and the formation of hot springs and geysers

igneous
rock formed by solidification of molten magma or lava

ignimbrite
deposit formed by an ash flow or nuée ardente eruption

intermediate
volcanic rock containing between 53-67% silica. Andesite and dacite are intermediate rocks

lahar
volcanic debris mobilised by the addition of water into a mudflow. Lahars may be caused in a number of ways e.g., the overflow or eruption of a crater lake, the bursting of a dam retaining a crater lake, heavy rain, or the eruption of hot material onto ice and snow

lapilli
small pyroclastic fragments of between 2-64mm in diameter formed when lava is ejected into the air by expanding gases

lava
molten rock (degassed magma) which flows from a vent onto the Earth's surface. The name also applies to the same material that has cooled and solidified

lava plug
lava which has solidified in the vent of a volcano, plugging it, like a cork in a bottle

magma
molten rock, including its volcanic gases, as it exists underground

mantle
the main bulk of the Earth, ranging from depths of about 40-2900km, between the crust and the core

massif
a block of mountains forming a single, compact connected group

moraine
ridges or deposits of rock debris transported by a glacier

nuée ardente
a French term applied to a turbulent and swiftly flowing mass of hot gases, ash and other pyroclastic material expelled with explosive force from a volcano

pH
a measure of the acidity or alkalinity of a solution. A pH of 7.0 is neutral; below 7.0 is acidic and above 7.0 is alkaline

phreatomagmatic
eruptions of this kind result from the conversion of ground water to steam by ascending magma. Fragments of lava are present in the ejecta

plates
the crust of the Earth is divided into a number of plates that are internally rigid and move independently over the mantle

plate tectonics
the theory of plate formation, movement, interaction and destruction which is used to explain mountain building and the distribution of earthquakes, as well as continental drift

pumice
a light-weight pyroclastic rock, with a porous (vesicular) structure, resulting from the expansion of volcanic gas

pyroclastic rocks
fragments of volcanic material, ranging widely in shape and size, that have been expelled from a volcano through the air

rhyolite
a type of volcanic rock with a high silica content (greater than 65%)

ring plain
circular apron of water-transported materials (lahar deposits and alluvium) surrounding a volcanic cone

scoria
fragments of lava, ejected through the air from a volcano, characterised by its rough appearance, dark colour and its vesicular nature

sedimentary rocks
rocks formed by the accumulation and cementation of material e.g., sand, silt and mud, derived from pre-existing rocks

seismic
of earthquake or earth vibrations

strombolian activity
semi- continuous , mild eruptive activity concentrated on the active vent. Fairly fluid lava is involved in these eruptions

subduction
the process by which one plate descends beneath another

subduction zone
a dipping planar zone descending away from a trench and defined by seismic activity between two plates

surtseyian eruptions
occur when large volumes of water get mixed up in a volcanic eruption, producing more violent eruptions than would otherwise take place

tectonics
the study of the movements and deformation of the crust of the Earth on a large scale, including folding, faulting and plate tectonics

tephra
a collective term for all unconsolidated pyroclastic ejecta

vent
an opening in the surface of the Earth's crust through which volcanic material (lava and pyroclastics) is forced during an eruption

viscosity
a measure of resistance to flow in a liquid i.e., a highly viscous lava does not flow easily

The peak of Paretetaitonga, "snow from the southern sea", can be seen rising above the ash-covered head of the Whangaehu Valley.

volcano
any vent through which volcanic materials are erupted, or the mountain or landform formed by volcanic eruptions

vulcanian eruptions
a type of eruption caused when gas pressure builds beneath an obstruction in the volcanic vent and results in violent explosive eruptions

welded airfall
separate fragments of pyroclastic material welded together by heat

Bibliography

COWAN, J. **The Tongariro National Park.** Tongariro National Park Board, Handbook, 1927.

FRANCIS, P. **Volcanoes.** Pelican Books, 1976.

GREGG, D.R. **Volcanoes of Tongariro National Park.** New Zealand DSIR Information Series, No. 28, 1960.

HACKETT, W.R. **Geology and Petrology of Ruapehu Volcano and Related Vents.** Unpublished PhD thesis, Victoria University of Wellington, 1985.

SOONS, J.M. AND SELBY, M.J. (Editors). **Landforms of New Zealand.** Longman Paul Ltd, 1982.

STEVENS, G. **New Zealand Adrift.** A.H. & A.W. Reed, 1980.

TOPPING, W.W. **Some aspects of Quaternary History of Tongariro Volcanic Centre.** Unpublished PhD thesis, held in Geology Library, Victoria University of Wellington, 1974.

Plus scientific papers and other publications too numerous to itemise which are listed in:

TURNBULL, L.H. **Bibliography for Tongariro National Park.** Department of Lands and Survey, 1979.

Also of interest: A geology teaching film prepared by David Branagan and photographed by John Lesnie. The Department of Geology and Geophysics, University of Sydney. **Ngauruhoe Erupts — The 1954 Eruption of Ngauruhoe, New Zealand.**

A rare event occured during the winter of 1926 – the surface of Ruapehu's usually warm crater lake froze – the only time it has done so in the last hundred years.
(Refer also pages 72 and 76)

INDEX

Acknowledgements

A book of this nature relies primarily on the knowledge and research of others and, thus, I would like to acknowledge the many people who have contributed to the wealth of information about Tongariro National Park. This began with the first explorers who travelled the region and recorded detailed observations through to the scientists of today who are still actively adding to the body of information about the Tongariro Volcanic Centre.

I am indebted to Dr Jim Cole of Canterbury University for his review of the manuscript for the first edition and to Dr Ian Nairn of the Institute of Geological and Nuclear Science who was involved with both the first and the fourth editions. Their assistance was invaluable and their ongoing research work is also acknowledged. Dr Bruce Houghton, formerly of IGNS, has generously helped me over the years and contributed to this edition. My grateful thanks to Dr Barbara Hobden for reviewing the Tongariro section and contributing her Ngauruhoe map on page 53. Volcanologist Dr Colin Wilson assisted by updating the section on the Taupo caldera.

The following specialists assisted with the first edition: Werner Giggenbach (Chemistry); Ray Deacon, Bill Hackett, Jim Healy, Bill McIntosh, Tim Stern, Colin Wilson (Geology); Arnold Heine, Harry Keys, Paul Robinson (Glaciology); Graham Bagnall (History); and Ray Dibble (Seismology).

A special thank you to my husband Dr Harry Keys for his sustained interest in this book which has continued to benefit from his broad-based knowledge of Tongariro National Park. I am also very grateful to photographer Lloyd Homer for his many brilliant photos. I acknowledge the contribution of Bruce Jefferies, former chief ranger of Tongariro National Park, who provided the impetus for this book and the Tongariro Natural History Society. Finally, I am indebted to the hard work and support of John and Pat Newton, Bob Stothart, Paul Green and other members of the Tongariro Natural History Society.

Karen Williams, April 2001.

Mt Ruapehu, 17 October 1995

Paintings

The following items have been reproduced with permission from the Alexander Turnbull Library: Chromolithograph of Lake Taupo and the Volcanoes by an unknown artist - p.6; Drawings of surveyors by Ferdinand von Hochstetter - p.34; Etching of Ngauruhoe by James Kerry-Nicholls - p.46; Watercolour of Ngauruhoe crater by Kennett Watkins - p.56.

Photographic Acknowledgements

Where more than one photo appears on a page, photos are credited in sequence starting from the left side, and working from top to bottom in each column.

Front cover:John King courtesy of Mountain Air, Inside cover: Communicate New Zealand Back Cover: Noel Woodfield

2:TNP 5&6:Bruce Jefferies 8:ATL 13:Neville Peat/Lloyd Homer/Jim Healy 16:Wendy St George 18:TNP 19:Jim Healy/Wendy St George/Wendy St George 21:Lloyd Homer/Lloyd Homer/*; 22:Bruce Jefferies 24:Lloyd Homer 25:Dave Bamford 29:TNP 30:Lloyd Homer 31:Bruce Jefferies 32:Lloyd Homer 33:Dave Bamford/Lloyd Homer 36&37:Lloyd Homer/Mike Edginton (inset) &38; 39&40&41:Bruce Jefferies 42:TNP/Bruce Jefferies 43&44:Lloyd Homer 45:Peter Simpson 46:ATL 48:TNP 49:Herb Spannagl 50:Graham Hancox/Lloyd Homer 52:Ted Lloyd/Jim Healy/Ted Lloyd 54:John Scobie 55:Jim Healy 56:ATL 57:Lloyd Homer/TNP 58:NZAM 60:Lloyd Homer 61:Bruce Jefferies 62:Barbara Cooper/Bruce Jefferies 63:Lloyd Homer 64:Dave Bamford/Lloyd Homer &65&66; 68:John Nankervis 69:Bill Hackett/Lloyd Homer (inset) &70; 72:ATL 73:APL 74&77:Lloyd Homer 78:Keith McNaughton 79:Bert Hollick/Bert Hollick/Peter Otway 80:Barbara Cooper/Jim Healy/Jim Healy 81:Bill Cooper/Lloyd Homer 82:Lottie McGrath/Horace Fyfe 83:TNP 84:IGNS 86:Royal NZ Air Force 87:Auckland Star 88&89:Lloyd Homer 90:Harry Keys 91&92:Lloyd Homer 94:John King 95:Jack Bedford; 96&97:TNP 98:TNP/Lloyd Homer 99:Bruce Jefferies 100:Dave Bamford/TNP 102:Barbara Cooper/TNP 103:TNP 104&106:Lloyd Homer 107:Lloyd Homer/Dave Bamford 108:TNP 111:Bill Cooper/Auckland Institute & Museum 112:APL/ATL 113:TNP/Karen Williams 114:Ron Keam 115:IGNS 116:Doug Sheppard 118:Michael Kopp/Lloyd Homer 119:Lloyd Homer 120:TNP/Victoria University 121:TNP 122:Thor Thordarson 123:Noel Woodfield 124:Bruce Houghton 125:Tim Whittaker 126:Nicola Topping - NZ Herald 127:Dave Rothschild 128:Barbara Curtis 129:Michael Rosen/Harry Keys/Peter Otway 130:Harry Keys 131:Manaaki Whenua Landcare Research 132:Alan Gibson - NZ Herald 133:Harry Keys 134:Bill Rackham 136:Harry Keys 137:Zinzuni Jurado-Chichay 138:Harry Keys 140&141:Wendy St George 143:John Nankervis 144:Ian Powell 146:Natural History New Zealand/J Hedley 148 John King courtesy of Mountain Air.

ATL - Alexander Turnbull Library
APL - Auckland Public Library
IGNS - Institute of Geological & Nuclear Sciences
TNP Tongariro National Park Collection, Whakapapa

* Original imagery obtained from NASA, USA and prepared by the Department of Conservation in cooperation with the Physics and Engineering Laboratory, DSIR.Mountain Air.

Published by:Tongariro Natural History Society, PO Box 238, Turangi, New Zealand.

First published 1985, second edition 1989, third edition 1994 and fourth edition 2001.
Pre-press: Scanner Graphic Ltd
Printing: Bellprint Ltd
Art Direction & Design: Leonard Cobb Direction Ltd.
Finished Art: Martina Cobb.

The young cone of Ngauruhoe rises above Pukekaikiore and other eroded western remnants of ‘proto Ngauruhoe’. Evening light highlights a young prehistoric lava flow in the head of the glaciated Makahikatoa Valley running down between two lateral moraine ridge systems.